7-11-89

Tissue Culture Techniques
for Horticultural Crops

Tissue Culture Techniques for Horticultural Crops

Kenneth C. Torres
Sigma Chemical Company
Tissue Culture Department
St. Louis, Missouri

An **avi** Book
Published by Van Nostrand Reinhold
New York

To Janet, Jacob, and Bradley

An AVI Book
(AVI is an imprint of Van Nostrand Reinhold)
Copyright © 1989 by Van Nostrand Reinhold

Library of Congress Catalog Card Number 88-5609

ISBN 0-442-28465-9

Printed in the United States of America

Van Nostrand Reinhold
115 Fifth Avenue
New York, New York 10003

Van Nostrand Reinhold (International) Limited
11 New Fetter Lane
London EC4P 4EE, England

Van Nostrand Reinhold
480 La Trobe Street
Melbourne, Victoria 3000, Australia

Macmillan of Canada
Division of Canada Publishing Corporation
164 Commander Boulevard
Agincourt, Ontario M1S 3C7, Canada

16 15 14 13 12 11 10 9 8 7 6 5 4 3 2 1

Library of Congress Cataloging-in-Publication Data

Torres, Kenneth C., 1957–
    Tissue culture techniques for horticultural crops.

    "An AVI book."
    Bibliography: p.
    Includes index.
    1. Horticultural crops—Propagation—In vitro.
2. Plant tissue culture. I. Title.
SB319.625.T67   1988   582'.00724   88-5609
ISBN 0-442-28465-9

Contents

Preface

This book was written for those individuals who are concerned about the techniques and practices of plant cell cultures for horticultural crops. It was designed to serve as a text and reference for students and professionals in ornamental horticulture, fruit and vegetable crop production, botany, forestry, and other areas of plant science.

Research during the last twenty-five years in the area of plant tissue culture has led to many developments and changes in this field. Although the techniques involved in the manipulation of plant tissue culture are now relatively straightforward, the presentation of these techniques in a short volume for the beginner in the field is generally unavailable. In addition to describing the techniques for establishment and manipulation of specific species, several chapters in this book also provide a brief, general review of important cultural parameters. Specific protocols and laboratory procedures may also be found in the appendix. I hope that this presentation of information will be helpful to those individuals wanting to apply plant tissue culture techniques for horticultural crops.

This book has been possible only with the encouragement and support of many people. I am deeply indebted to my wife, Janet, for her patience and understanding, her editing, and her constant encouragement. This manuscript could not have been completed without her support. My deepest appreciation goes to Dr. Donald W. Newsom for his encouragement to begin writing the manuscript and to my graduate students who kept me constantly thinking by asking perceptive questions and helping with the photography and illustrations found within the book. Many thanks to my colleagues at Sigma Chemical Company for their encouragement with this project.

PART I

Basic Techniques and Principles

1

Overview of Facilities and Techniques

The term *plant tissue culture* broadly refers to the cultivation *in vitro* of all plant parts (single cells, tissues, and organs) under aseptic conditions. Plant tissue culture systems are often used as "model" systems in the study of various physiological, biochemical, genetic, and structural problems related to plants. Plant tissue culture techniques also have great potential as a means of vegetatively propagating economically important crops and crops of future potential on a commercial basis (2).

The techniques of plant tissue, cell, and organ culture are now well established in many research laboratories throughout the world. Methods have been developed for the generation and screening of desirable variants; cellular cloning and rapid propagation of genotypes; induction of haploid tissue from anther and pollen cultures; extending the range of genetic variability by way of induced mutations and somatic clones; and formation of isolated callus and cell cultures for studies on the effects of nutrients, vitamins, and hormones on cell growth and differentiation. Many of the specific tissue culture techniques being used at present are described in later chapters. The future of tissue culture rests in its use as a tool in basic and applied research and the subsequent utilization of these techniques for commercial production.

Although the term plant tissue culture is commonly used to include all types of aseptic plant culture, it is sometimes preferable to use the following more specific terms to distinguish the various types of culture:

- Plant culture—culture of seedlings or larger plants
- Embryo culture—culture of isolated mature or immature embryos
- Organ culture—culture of isolated plant organs
- Tissue or callus culture—culture of tissue arising from explants of plant organs
- Suspension culture and cell culture—culture of isolated cells or very small cell aggregates remaining dispersed in liquid medium

- Protoplast culture—culture of plant protoplasts, i.e., cells devoid of their retaining walls
- Anther or haploid culture—culture of anthers and/or immature pollen grain in an effort to obtain a haploid cell or callus line.

All types of plant tissue culture, both old and new, involve two fundamental steps (1). First, the plant part or explant must be isolated from the rest of the plant body. This disrupts the cellular, tissue, and/or organ interactions that may occur in the intact plant. Second, the excised plant part must be placed in an appropriate environment in which it can express its intrinsic or induced potential. Both the chemical composition of the medium and the physical conditions of the environment (e.g., gaseous atmosphere, type of culture vessel, light and temperature conditions) should effectively control the expression of any genotypic or phenotypic potential in the explant. Both of these steps must be carried out aseptically; that is, the culture must be free of all bacterial, fungal, and other contaminants because these may cause either the overgrowth of the explant or the production of metabolites that may be toxic or influence the explant's metabolic growth, and/or developmental processes.

The need to prevent contamination by microorganisms determines, in large part, the general organization and many of the procedures used in a tissue culture laboratory. In this chapter, the basic organization of a tissue culture facility and common experimental procedures used in nearly all tissue culture work are described. The preparation of culture media, which is critical to successful tissue culture, is discussed in Chapter 2.

ORGANIZATION OF A TISSUE CULTURE LABORATORY

Any laboratory in which tissue culture techniques are performed, regardless of the specific purpose, must contain a number of basic facilities (2). These usually include the following:

- A general washing area
- A media preparation, sterilization, and storage area
- An aseptic transfer area
- Environmentally controlled incubators or culture rooms
- An observation/data collection area

The design of the latter area depends upon the nature of research undertaken and the method of evaluation. The layout of a typical tissue culture facility is shown in Fig. 1.1.

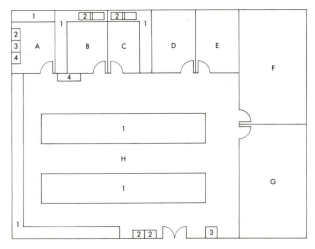

Fig. 1.1. Diagram of typical tissue culture facility. A—Sterilization room with lab bench (1), autoclaves (2, 3), and heat (dry) sterilization unit (4). B—Washing area with lab bench (1) and double-basin, lead-lined sink (2). C—Darkroom with lab bench (1) and double-basin, lead-lined sink (2). D—Storage room for chemicals and glassware. E—Transfer room. F and G—Temperature-controlled culture rooms. H—Main laboratory with lab benches and tables (1), refrigerators (2), centrifuge (3), and water distillation station (4) with a feed to washing area (B).

Washing Area

The washing area should contain large sinks, some lead-lined to resist acids and alkalis, draining boards, and racks, and have access to demineralized water, distilled water, and double-distilled water. Space for drying ovens or racks, automated dishwashers, acid baths, pipette washers and driers, and storage cabinets should also be available in the washing area.

Media Preparation Area

The media preparation area should have ample storage space for the chemicals, culture vessels and closures, and glassware required for media preparation and dispensing. Bench space for hot plates/stirrers (Fig. 1.10), pH meters, balances, water baths, and media-dispensing equipment should be available. Other necessary equipment may include air and vacuum sources, distilled and double-distilled water, bunsen burners with a gas source, refrigerators and freezers for storing stock solutions and chemicals, a microwave or a convection oven, and an autoclave or domestic pressure cooker for sterilizing media, glassware, and instruments.

In preparing culture media, analytical grade chemicals should be used and good weighing habits practiced. To insure accuracy, an exact step-by-step routine should be developed for media preparation and a complete checklist required of all media preparers even for the simplest media. Media formulation and procedures are described in Chapter 2 and summarized in Appendix 4.

The water used in preparing media must be of the upmost purity and highest quality. Tap water is unsuitable because it may contain cations (ammonium, calcium, iron, magnesium, sodium, etc.), anions (bicarbonates, chlorides, fluorides, phosphates, etc.), microorganisms (algae, fungi, bacteria), gases (oxygen, carbon dioxide, nitrogen), and particulate matter (silt, oils, organic matter, etc.) (5). Water used for plant tissue culture should meet, at a minimum, the standards for type II reagent grade water, i.e., be free of pyrogens, gases, and organic matter and have an electrical conductivity less than 1.0 μmho/cm (2, 4).

The most common and preferred method of purifying water to type II standards is a deionization treatment followed by one or two glass distillations (Fig. 1.2). The deionization treatment removes most ionic impurities, and the distillation process removes large organic molecules, microorganisms, and pyrogens (2). Three other methods that will produce type II purity water are (1) absorption filtration, which uses activated carbon to remove organic contaminants and free chlorine; (2) membrane filtration, which removes particulate matter and most bacterial contamination; and (3) reverse osmosis, which removes approximately 99% of the bacterial, organic, and particulate matter as well as about 90% of the ionized impurities (2).

Transfer Area

Under very clean and dry conditions, tissue culture techniques can be successfully performed on an open laboratory bench. However, it is advisable that a laminar flow hood or sterile transfer room be utilized for making transfers. Within the transfer area there should be a source of electricity, gas, compressed air, and vacuum.

The most desirable arrangement is a small dust-free room equipped with an overhead ultraviolet light and a positive-pressure ventilation unit. The ventilation should be equipped with a high-efficiency particulate air (HEPA) filter. A 0.3-μm HEPA filter of 99.97−99.99% efficiency works well. All surfaces in the room should be designed and constructed in such a manner that dust and microorganisms do not accumulate and the surfaces can be thoroughly cleaned and disinfected. A room of such design is particularly useful if large numbers of

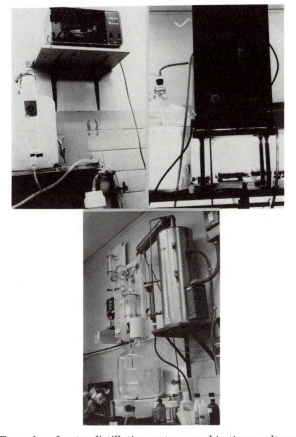

Fig. 1.2. Examples of water distillation systems used in tissue culture facilities.

cultures are being manipulated or large pieces of equipment are being utilized.

Another type of transfer area is a laminar flow hood (Fig. 1.3). Air is forced into the unit through a dust filter then passed through a HEPA filter. The air is then either directed downward (vertical flow unit) or outward (horizontal flow unit) over the working surface. The constant flow of bacteria-free filtered air prevents nonfiltered air and particulate matter from settling on the working surface.

The simplest type of transfer area suitable for tissue culture work is an enclosed plastic box commonly called a glove box. This type of culture hood is sterilized by an ultraviolet light and wiped down peri-

Fig. 1.3. Laminar flow hood used for transferring materials under sterile conditions.

odically with 95% ethyl alcohol when in use. This type of unit is used when relatively few transfers are required.

Culture Room

All types of tissue cultures should be incubated under conditions of well-controlled temperature, humidity, air circulation, and light quality and duration. These environmental factors may influence the growth and differentiation process directly during culture or indirectly by affecting their response in subsequent generations. Protoplast cultures, low-density cell suspension cultures, and anther cultures are all particularly sensitive to environmental cultural conditions (3, 6, 7, 8).

Typically, the culture room for growth of plant tissue cultures should have a temperature between 15° and 30°C, with a temperature fluctuation of less than ±0.5°C; however, a wider range in temperature may be required for specific experiments. It is also recommended that the room have an alarm system to indicate when the temperature has reached preset high or low temperature limits, as well as a continuous temperature recorder to monitor temperature fluctuations. The temperature should be constant throughout the entire culture room (i.e., no hot or cold spots). The culture room should have enough fluorescent lighting to reach 10,000 lux; the lighting should be adjustable in terms of quantity

Fig. 1.4. Cultures must be incubated in a controlled environment such as a walk-in growth chamber (top left); an upright, refrigerator-type incubator (top right); or a reach-in incubator (bottom).

and photoperiod duration. Both light and temperature should be programmable for a 24-hr period. The culture room should have fairly uniform forced-air ventilation, and a humidity range of 20–98% controllable to ±3%. Many incubators, large growth chambers, and walk-in environmental chambers meet these specifications (Figs. 1.4 and 1.5).

Fig. 1.5.   Shakers are required for maintaining cell suspension cultures. Shakers may be placed in a temperature-controlled culture room (top and middle) or in a temperature-controlled water bath (bottom).

BASIC LABORATORY EQUIPMENT

Many tissue culture techniques require similar basic laboratory equipment. The following items are commonly found in a laboratory for *in vitro* propagation of plant materials:

1. Hot plate or small stove
2. Glass or stainless steel containers for heating and dissolving media
3. Pressure steam sterilizer
4. pH meter
5. Centigram balance
6. Graduated measuring cylinders
7. Culture tubes, bottles, and other glassware with suitable closures
8. Dispensing devices
9. Small transfer instruments (e.g., spatulas, scalpels, and forceps)
10. Refrigerator
11. Water deionizer or source of deionized water
12. Disinfectants
13. Chemicals for culture media or commercially prepared culture media
14. Illuminated magnifier or stereomicroscope (Fig. 1.6)

The glassware used in tissue culture can generally be found in most laboratories (Fig. 1.7). The glassware, particularly the culture vessels, should be made of Pyrex or boro-silicate glass. Due to the increasing expense of this type of glass, many laboratories are successfully converting to soda glass, which may be seven to eight times cheaper. Wide-neck Erlenmeyer flasks (50-, 125-, 250-ml capacity) are commonly used as culture vessels; large volume Erlenmeyer flasks are required for media preparation. Test tubes, petri dishes, mason jars, baby food jars, and other glassware can also be adapted to tissue culture. Since all new glass may release substances that affect the composition of the medium, it is recommended that all new glassware be filled with water, autoclaved twice with detergent washes, and rinsed between washes before being used for tissue culture. Other glassware commonly required in a tissue culture facility includes beakers, volumetric flasks, pipettes, and graduated cylinders.

Directions for operating a water bath, water still, pH meter, Oxford automatic dispenser, Drummond Pipet-Aid, and shakers are given in Appendices 13, 15, 16, and 18.

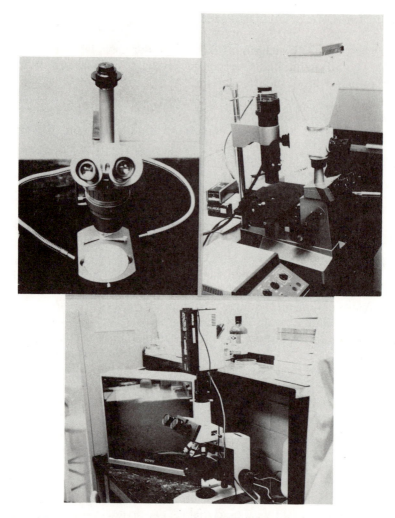

Fig. 1.6. Microscopes may be required for certain tissue culture procedures. A dissecting microscope (top left) is commonly used for removing shoot tips and apical domes from stems; an inverted microscope (top right) is used for studying protoplasts and cell suspension cultures. A compound microscope (bottom) is used for observing prepared microscope slides, making cell counts, etc.

BASIC LABORATORY PROCEDURES

The majority of laboratory operations utilized in the *in vitro* propagation of plants can be learned easily. One needs to concentrate mainly on accuracy, cleanliness, and strict adherence to details when performing *in vitro* techniques.

Fig. 1.7. Various types of culture vessels commonly used for culturing plant cells (top) and common laboratory glassware used in media preparation (bottom).

Weighing

The preparation of media requires careful weighing of all components. Even if a commercially prepared medium is used, care must be taken in preparing it and any stock solutions that are required. Table 1.1 lists common units of dry measure and their relative value.

Because of the diversity of laboratory balances in use, it is impossible to review the details of their operation. The manufacturer's instructions should be consulted before using any balance. The types of balances most often encountered in the laboratory include top-loading single-pan balance, triple-beam balance, double-pan torsion balance, analytical single-pan balance, and top-loading electronic balance (Fig. 1.8). The last type has become quite popular in recent years due to its accuracy, ease of use, and durability. With certain models of top-loading

Table 1.1. Common Units of Dry and Liquid Measure[a]

Unit	Symbol	Relative Value
Dry Measure		
Kilogram	kg	1000 grams
Gram	g	1 gram
Centigram	cm	0.01 gram
Milligram	mg	0.001 gram
Microgram	μg	0.000001 gram
Ounce	oz	28.3 grams
Pound	lb	454 grams
Liquid Measure		
Liter	l	1 liter
Milliliter	ml	0.001 liter
Microliter	μl	0.000001 liter
Quart	qt	0.946 liter
Ounce	fl oz	29.6 ml
1 ml = 1 cubic centimeter (cm^3 or cc)		

[a] See Appendix 2 for interconversion of some common units of measure.

electronic balances, milligram accuracy is possible; in the past such accuracy required the use of analytical balances. General instructions for operating a top-loading pan and Mettler analytical balance are presented in Appendix 14.

Several common precautions must be observed to obtain accurate weights. First, the balance should be located on a hard, stable, level surface which is free of vibrations and excessive air drafts. The balance or weigh area should always be kept clean. Most importantly, the balance should *never* be overloaded (see manufacturer's specification). It is advisable to use a lightweight weighing container or paper rather than placing the material to be weighed directly on the pan surface.

Measuring Liquids

Calibrated glassware (e.g., beakers, flasks, and pipettes) are required for the preparation of culture media. Graduated cylinders of 10-, 25-, 100-, and 1000-ml capacities are used for many measuring operations, but volumetric flasks and pipettes (Fig. 1.9) are required for more precise measurements. Measurement of solutions with pipettes or graduated cylinders is only accurate when the bottom of the curved air–liquid interface is aligned with the measuring mark. Common units of liquid measure are listed in Table 1.1.

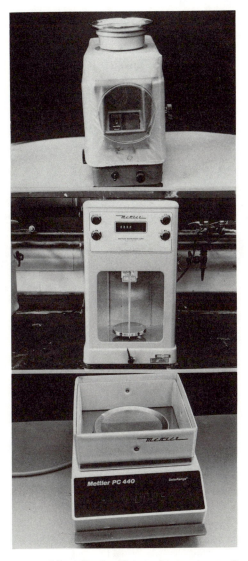

Fig. 1.8. Balances are used for collecting data and preparing media. A top-loading pan balance (top) or top-loading electronic balance (bottom) are best suited for media preparation; an analytical (middle) or top-loading electronic (bottom) balance should be used for data collection.

Pipettes should be filled with a hand-operated device, called a pipettor, which eliminates the hazards of pipetting by mouth. **Never pipette by mouth!!** Three types of pipettors are commonly used. The first is a bulb-type pipettor, which is controlled by a series of valves. The second type of pipettor is operated simply by rotating a small wheel on the

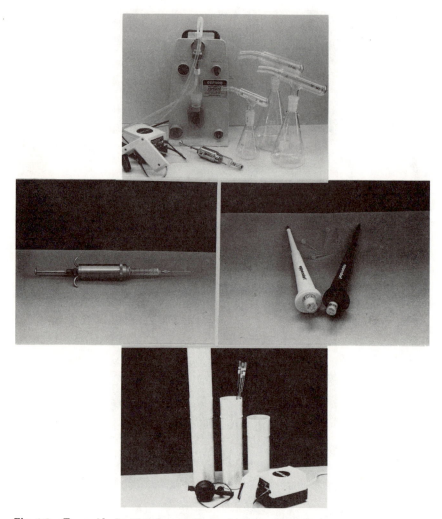

Fig. 1.9. To avoid pipetting by mouth, various types of pipettors are used in tissue culture work. Illustrated are several pipettors for dispensing media (top) and for disrupting cells and measuring small quantities (middle). Devices for washing and sterilizing pipettes (bottom) also are needed.

side of the handle, which creates a suction bringing the liquid up the pipette; rotating the wheel in the opposite direction releases the suction. A third type of pipettor utilizes an electric air pump. Liquid is drawn into the pipette by pressing the top button and released by pressing the lower button. Directions for operating a Drummond Pipet-Aid, an example of the third type of pipettor, are given in Appendix 16.

Cleaning Glassware

The conventional method of washing glassware involves soaking the glass in a chromic acid–sulfuric acid bath followed by tap water rinses, distilled water rinses, and finally double-distilled water rinses (6). Due to the corrosive nature of chromic acid, the use of this procedure has been eliminated except for highly contaminated or soiled glassware. Adequate cleaning of most glassware for tissue culture purposes can be achieved by washing in hot water (70°C+) with commercial detergents, rinsing with hot tap water (70°C+), and finally rinsing with distilled and double-distilled water. However, highly contaminated glassware should be cleaned in a chromic acid–sulfuric acid bath or by some other proven method such as (1) ultrasonic cleaning, (2) washing with sodium pyrophosphate, or (3) boiling in metaphosphate (alconox), rinsing then boiling in a dilute hydrochloric acid solution, and then finally rerinsing. Cleaned glassware should be inspected, dried at 150°C in a drying oven, capped with aluminum foil, and stored in a closed cabinet.

The following general procedure is recommended for cleaning glassware that contains media and cultures after all data have been collected:

1. Autoclave all glassware, with media and cultures still in it, to kill any contaminating microorganisms that may be present.
2. After autoclaving the vessels, pour the media, after it has cooled but while it is still in a liquid state, into bio-hazard plastic bags or thick plastic bags, seal, then discard.
3. Wash all glassware in hot soapy water, using a suitable bottle brush to clean the internal parts of the glassware. Any glassware that is stained should be soaked in a concentrated sulfuric acid–potassium dichromate acid bath for 4 hr, then rinsed 10 times before washing with soapy water.
4. All glassware should be rinsed three times in tap water, three times in deionized water, and three times in double-distilled water, then dried and stored in a clean place.
5. Wash all instruments and new glassware in a similar manner.

Additional information about and a summary of cleaning procedures are presented in Appendix 3.

Sterilization

The most tedious parts of *in vitro* techniques are sterilizing plant materials and media and maintaining aseptic conditions once they have been achieved. Bacteria and fungi are the two most common contaminants observed in cell cultures. Fungal spores are lightweight and present throughout our environment. When a fungal spore comes into contact with the culture media used in tissue culture, conditions are optimal for germination of the spore and subsequent contamination of the culture.

Sterilizing Culture Rooms and Transfer Hoods

Large transfer rooms are best sterilized by exposure to ultraviolet (UV) light. Sterilization time varies according to the size of the room and should only be done when there are no experiments in progress. **Ultraviolet radiation is harmful to the eyes.** Transfer rooms can also be sterilized by washing them 1–2 times a month with a commercial brand of antifungal spirocyte. Smaller transfer rooms and hoods also can be sterilized with UV lights or by treatment with bactericides and/ or fungicides. Laminar flow hoods are easily sterilized by turning on the hood and wiping down all surfaces with 95% ethyl alcohol 15 min before initiating any operation under the hood.

Culture rooms should be initially cleaned with a detergent-brand soap, then carefully wiped down with a 2% sodium hypochlorite solution or 95% ethyl alcohol. All floors and walls should be washed gently on a weekly basis with a similar solution; extreme care must be used to avoid stirring up any contamination that has settled. Commercial disinfectants such as Lysol, Zephiram, and Roccal diluted at manufacturer's recommended rates can be used to disinfect work surfaces and culture rooms.

Sterilizing Glassware and Instruments

Metal instruments, glassware, aluminum foil, etc., can all be sterilized by exposure to hot dry air (130°–170°C) for 2–4 hr in a hot-air oven. All items should be sealed before sterilization but not in paper, as it decomposes at 170°C. Autoclaving is not advisable for metal instruments because they may rust and become blunt under these conditions.

Before instruments that have been sterilized in hot dry air are used, they should be removed from their wrapping, dipped in 95% ethyl alcohol, and exposed to the heat of a flame (Fig. 1.10). After an instrument has been used, it can again be dipped in ethyl alcohol, reflamed, and then reused. This technique is called *flame sterilization*. Safety is a major concern when using ethyl alcohol. Alcohol is *flammable* and if spilled near a flame will cause an instant flash fire. This problem is compounded in laminar flow hoods due to the strong air currents blown towards the worker. Fires commonly start when a flamed instrument is thrown back into the alcohol beaker. **In case of fire do not panic.** Fires can easily be put out by limiting the supply of oxygen.

Autoclaving is a method of sterilizing with water vapor under pressure. Cotton plugs, gauze, labware, plastic caps, glassware, filters, pi-

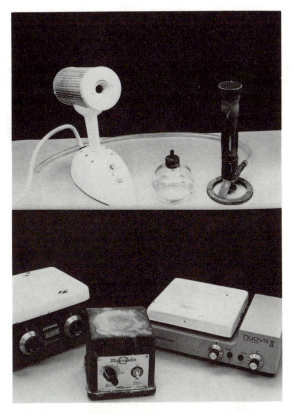

Fig. 1.10. Instruments must be sterilized before they are used to handle explants. The most common method is to soak an instrument in ethanol and then flame it using one of the burners shown (top). Stirrers and/or hot plates (bottom) are required for media preparation.

pettes, water, and nutrient media can all be sterilized by autoclaving. Nearly all microbes are killed by exposure to the super-heated steam of an autoclave for 10–15 min. All objects should be sterilized at 121°C and 15 psi for 15–20 min.

Sterilizing Nutrient Media

Two methods—autoclaving and membrane filtration under positive pressure—are commonly used to sterilize culture media. Culture media, distilled water, and other stable mixtures can be autoclaved in glass containers that are sealed with cotton plugs, aluminum foil, or plastic closures. However, solutions that contain heat-labile components must be filter-sterilized.

Generally, nutrient media are autoclaved at 15 psi and 121°C (Fig. 1.11). For small volumes of liquids (100 ml or less), the time required

Fig. 1.11. Large quantities of media generally are sterilized by heat using an autoclave similar to one of those shown.

for autoclaving is 15–20 min, but for larger quantities (2–4 liters), 30–40 min is required (Appendix 4). The pressure should not exceed 20 psi, as higher pressures may lead to the decomposition of carbohydrates and other thermolabile components of a medium.

Since many proteins, vitamins, amino acids, plant extracts, hormones, and carbohydrates are thermolabile and may decompose during autoclaving, filter sterilization may be required (Fig. 1.12). A Millipore or Seitz filter can be used; the porosity of the filter membrane should be no larger than 0.2 microns (μm). Empty glassware that is to hold media must be sterilized in an autoclave before filter sterilization. Additional information on sterile filtration is presented in Appendix 8, and operation of a Seitz filter is described in Appendix 17.

Nutrient media that contain thermolabile components can be prepared in several steps. That is, a solution of the heat-stable components is sterilized in the usual way by autoclaving, then cooled to 50°–60°C

Fig. 1.12. Media and media components may be sterilized using a filter sterilization unit similar to one of those shown. Heat-labile materials may require filter sterilization because they may decompose during normal autoclaving.

under sterile conditions; in a separate operation, solutions of the thermolabile components are filter-sterilized. The sterilized solutions are then combined under aseptic conditions to give the complete media.

Sterilizing Plant Material

Obtaining sterile plant material is difficult, and despite any precautions taken, 95% of cultures will end up contaminated if the explant is not disinfected in some manner. Because living materials cannot be exposed to extreme heat and retain their biological capabilities, plant organs and tissues are sterilized by treatment with a disinfecting solution. The most widely used solutions are listed in Table 1.2, along with the exposure times required to achieve sterilization. Solutions used to sterilize explants must preserve the plant tissue but at the same time destroy any fungal or bacterial contaminants.

Once explants have been obtained, they should be washed in a mild soapy detergent before treatment with a sterilizing solution. Some herbaceous plant materials (e.g., African violet leaves) may not require this step, but woody material, tubers, etc., must be washed thoroughly. After the tissue is washed, it should be rinsed under running tap water for 10–30 min and then be submerged into the disinfectant under sterile conditions. All surfaces must be in contact with the sterilant. After the allotted time for sterilization, the sterilant should be decanted and the explants washed at least three times in sterile distilled water. For materials that are difficult to disinfect, it may be necessary to repeat the treatment 24–48 hr before making the final explants. This allows previously unkilled microbes time to develop to a stage at which they are

Table 1.2. Commonly Used Disinfectants for Sterilizing Explants

Disinfectant	Concentration	Exposure Time (min)
Calcium hypochlorite	9–10%	5–30
Sodium hypochlorite[a]	0.5–5%	5–30
Hydrogen peroxide	3–12%	5–15
Ethyl alcohol	75–95%	[b]
Silver nitrate	1%	5–30
Bromine water	1–2%	2–10
Mercuric chloride	0.1–1.0%	2–10
Antibiotics	4–50 mg/liter	30–60
Benzalkonium chloride[c]	0.01–0.1%	5–20

[a] Commercial bleach contains about 5% sodium hypochlorite and thus may be used at 10–20% v/v concentration.
[b] Several seconds to several minutes.
[c] Zephiran, BTC, or Roccal.

vulnerable to the sterilant. Additional information on sterilizing plant material is presented in Chapter 3.

Sterile Culture Techniques

Successful control of contamination depends largely upon the operator's techniques in aseptic culture. You should always be aware of potential sources of contamination such as dust, hair, hands, and clothes. Obviously, your hands should be washed, sleeves rolled up, long hair tied back, etc. Washing your hands with 95% ethyl alcohol is an easy precautionary measure that can be taken. Talking or sneezing while culture material is exposed also can lead to contamination. When transferring plant parts from one container to another, do not touch the inside edges of either vessel. By observing where contamination arises in a culture vessel, you may be able to determine the source of contamination.

Determining pH

The pH of a solution is a measure of the concentration of hydrogen ions in the solution. The pH scale extends from very acid (0) to very alkaline (14) with 7 being the "neutral" point. The pH of most culture media is adjusted to 5.7 ± 0.1 before autoclaving. The pH can influence the solubility of ions in nutrient media, the ability of agar to gel, and the subsequent growth of cells. Thus, accurate determination and control of the pH of tissue culture media are necessary. Most commonly, pH is determined with a pH meter (Fig. 1.13). Instructions for operating a standard pH meter are given in Appendix 15.

LABORATORY SAFETY AND DAILY MAINTENANCE OPERATIONS

Observance of commonsense safety practices can significantly reduce the possibility of accidents or injuries occurring in a laboratory. For your safety and that of others, observe the following:

- Always wear shoes and a laboratory jacket in the laboratory.
- Be extremely careful handling alcohols around open flames. They are flammable!
- Never pipette by mouth.
- Handle hydrochloric acid, sulfuric acid, sodium hydroxide, and other strong acids and alkalis with extreme caution. They are very corrosive!
- Wash and bandage all cuts immediately.

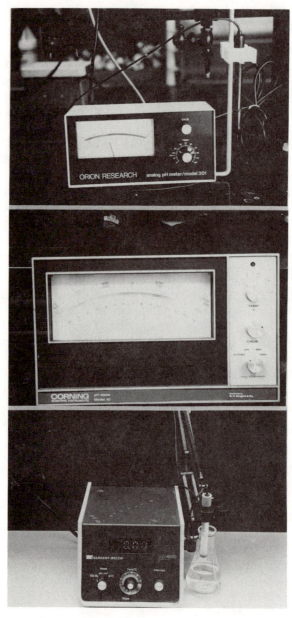

Fig. 1.13. Numerous types of pH meters are available for use in adjusting the pH of culture media.

• Before opening an autoclave, be sure the pressure is reduced to zero and the temperature is below 100°C.

In addition to safety concerns, cleanliness and proper care of equipment are vital to the operation of a tissue culture laboratory. The following tasks should be performed routinely before the laboratory is closed at night:

1. Mop floor in lab and culture room with an approved disinfectant.
2. Turn off hood, unless used continuously to reduce particulates in the air.
3. Fill distilled water tanks and turn off stills.
4. Clean off benches *completely* and put away chemicals, instruments, glassware, etc.
5. Put microscopes on lowest magnification; turn off and cover them.
6. Shut off all water outlets.
7. Set out all photographic negatives to dry.
8. Wash and dry all dirty glassware (tubes, pipettes, flasks, etc.).
9. Put away all clean, dry glassware, racks, etc.
10. Turn off all electrical equipment that is not in use overnight (e.g., stirrers, pH meters, balances).
11. Put away all photographic and/or chemical solutions.

REFERENCES

1. Biondi, S. and T. A. Thorpe. 1981. Requirements for a tissue culture facility. pp. 1–20. *In* Plant Tissue Culture: Methods and Applications in Agriculture. T. A. Thorpe (editor). Academic Press, New York.
2. Brown, D. C. W. and T. A. Thorpe. 1984. Organization of a plant tissue culture laboratory. pp. 1–12. *In* Cell Culture and Somatic Cell Genetics of Plants, Vol. 1. D. K. Vasil (editor). Academic Press, New York.
3. Fujiwara, A. 1982. Plant Tissue Culture, 1982. Maruzen, Tokyo.
4. Meltzer, R. L. 1979. Annual Book of ASTM Standards, Part 31—Water. Am. Soc. Test. Mater., Philadelphia, PA.
5. Pumper, R. W. 1973. Purification and standardization of water for tissue culture. pp. 674–677. *In* Tissue Culture Methods and Applications. P. F. Kruse, Jr., and K. M. Patterson (editors). Academic Press, New York.
6. Street, H. E. 1973. Laboratory organization. pp. 11–30. *In* Plant Tissue and Cell Culture. H. E. Street (editor). Univ. of California Press, Berkeley.
7. Street, H. E. (editor). 1974. Tissue Culture and Plant Science. Academic Press, New York.
8. Thorpe, T. A. (editor). 1978. Frontiers of Plant Tissue Culture, 1978. Univ. of Calgary Press, Calgary, Canada.
9. White, P. R. 1941. Plant tissue cultures. Biol. Rev. Cambridge Philos. Soc. 16:34–48.

2

Tissue Culture Media—Composition and Preparation

MEDIA COMPONENTS

One of the most important factors governing the growth and morphogenesis of plant tissues in culture is the composition of the culture medium. The basic nutrient requirements of cultured plant cells are very similar to those of whole plants. The historical development of plant tissue culture media is described in numerous review articles and books and will not be discussed herein (27, 33, 67, 70).

Plant tissue and cell culture media are generally made up of some or all of the following components: macronutrients, micronutrients, vitamins, amino acids or other nitrogen supplements, sugar(s), other undefined organic supplements, solidifying agents or support systems, and growth regulators. Several media formulations are commonly used for the majority of all cell and tissue culture work. These media formulations include those described by White (84), Murashige and Skoog (53), Gamborg et. al. (26), Gautheret (30), Schenk and Hilderbrandt (60), Nitsch and Nitsch (55), and Lloyd and McCown (44, 45, 48). Murashige and Skoog's MS medium, Schenk and Hilderbrandt's SH medium, and Gamborg's B-5 medium are all high in macronutrients, while the other media formulations contain considerably less of the macronutrients.

Macronutrients

The macronutrients provide the six major elements—nitrogen (N), phosphorus (P), potassium (K), calcium (Ca), magnesium (Mg), and sulfur (S)—required for plant cell or tissue growth. The optimum concentration of each nutrient for achieving maximum growth rates varies considerably among species.

Culture media should contain at least 25–60 mM of inorganic nitrogen for adequate plant cell growth. Plant cells may grow on nitrates alone, but considerably better results are obtained when the medium contains both a nitrate and ammonium nitrogen source. Certain species require ammonium or another source of reduced nitrogen for cell growth to occur. Nitrates are usually supplied in the range of 25–40 mM; typical ammonium concentrations range between 2 and 20 mM. However, ammonium concentrations in excess of 8 mM may be deleterious to cell growth of certain species. Cells can grow on a culture medium containing ammonium as the sole nitrogen source if one or more of the TCA cycle acids (e.g., citrate, succinate, or malate) are also included in the culture medium at concentrations of approximately 10 mM (18). When nitrate and ammonium sources of nitrogen are utilized together in the culture medium, the ammonium ions will be utilized more rapidly and before the nitrate ions.

Potassium is required for cell growth of most plant species. Most media contain K, in the nitrate or chloride form, at concentrations of 20–30 mM. The optimum concentrations of P, Mg, S, and Ca range from 1–3 mM when all other requirements for cell growth are satisfied. Higher concentrations of these nutrients may be required if deficiencies in other nutrients exist.

Micronutrients

The essential micronutrients for plant cell and tissue growth include iron (Fe), manganese (Mn), zinc (Zn), boron (B), copper (Cu), and molybdenum (Mo). Chelated forms of iron and zinc are commonly used in preparing culture media. Iron may be the most critical of all the micronutrients. Iron citrate and tartrate may be used in culture media, but these compounds are difficult to dissolve and frequently precipitate after media are prepared. Murashige and Skoog (53) used an ethylene diaminetetraacetic acid (EDTA)–iron chelate to bypass this problem. Nitsch (54) demonstrated that the chelated form was superior to iron citrate for embryo induction. EDTA chelates are not entirely stable, particularly in liquid culture, and may precipitate after a few days (67). Steiner and van Winden (65) have developed a procedure for preparing an iron chelate solution that will not precipitate.

Cobalt (Co) and iodine (I) may also be added to certain media, but strict cell growth requirements for these elements have not been established. Sodium (Na) and chlorine (Cl) are also used in some media but are not essential for cell growth. Copper and Co are normally added to culture media at concentrations of 0.1 μM, Fe and Mo at 1 μM, I at 5 μM, Zn at 5–30 μM, Mn at 20–90 μM, and B at 25–100 μM.

Carbon and Energy Source

The preferred carbohydrate in plant cell culture media is sucrose. Glucose and fructose may be substituted in some cases, glucose being as effective as sucrose and fructose being somewhat less effective. Other carbohydrates that have been tested include lactose, galactose, raffinose, maltose, and starch, but these carbohydrates all generally produce results inferior to those obtained with either sucrose or glucose. Thorpe (72) has shown that in tobacco at least part of the sucrose requirement is osmotic. He noted that when sucrose concentrations were lowered and mannitol substituted to give an equivalent weight-to-volume osmoticum, there was no reduction in shoot production. Sucrose concentrations of culture media normally range between 2 and 3%.

Carbohydrates must be supplied to the culture medium because very few plant cell lines have been isolated that are fully autotropic, e.g., capable of supplying their own carbohydrate needs by CO_2 assimilation during photosynthesis (6, 11, 42). The sucrose present in the culture medium is rapidly broken down to fructose and glucose. Glucose is utilized by the cells first, followed by fructose. A partial hydrolysis of sucrose occurs when media are autoclaved (5). The extent of this hydrolysis is greater when sucrose is autoclaved in the presence of other medium components than when it is autoclaved alone (21). Cultures of some plant species grow better on media with autoclaved sucrose than on media with filter-sterilized sucrose, suggesting that the cells benefit from a ready supply of glucose and fructose (5). Use of autoclaved fructose can be detrimental to cell growth.

Vitamins

Normal plants synthesize the vitamins required for their growth and development. Vitamins are required by plants as catalysts in various metabolic processes. When plant cells and tissues are grown *in vitro*, some vitamins may become limiting factors for cell growth. The vitamins most frequently used in cell and tissue culture media include thiamin (B_1), nicotinic acid, pyridoxine (B_6), and myo-inositol. Thiamin is the one vitamin that is basically required by all cells for growth (56). Discovery of the requirement for vitamin B_1 was simultaneously reported by Bonner (8, 9), Robbins and Bartley (58), and White (81). Thiamin is normally used at concentrations ranging from 0.1 to 10.0 mg/liter. Nicotinic acid and pyridoxine are often added to culture media (10, 30, 82) but are not essential for cell growth in many species. Nicotinic acid is normally used at concentrations of 0.1–5.0 mg/liter; pyridoxine is used at 0.1–10.0 mg/liter.

Myo-inositol is commonly included in many vitamin stock solutions. Although it is a carbohydrate not a vitamin, it has been shown to stimulate growth in certain cell cultures. Its presence in the culture medium is not essential, but in small quantities myo-inositol stimulates cell growth in most species (52). Myo-inositol is believed to be broken down into ascorbic acid and pectin and is incorporated into phosphoinositides and phosphatidylinositol, which play a role in cell division. Myo-inositol is generally used in plant cell and tissue culture media at concentrations of 50–5000 mg/liter.

Other vitamins such as biotin, folic acid, ascorbic acid, pantothenic acid, vitamin E (tocopherol), riboflavin, and p-aminobenzoic acid have been included in some cell culture media. The requirement for these vitamins by plant cell cultures is generally negligible, and they are not considered growth-limiting factors. These vitamins are generally added to the culture medium only when the concentration of thiamin is below the desired level or when it is desirable to grow cells at very low population densities.

Amino Acids or Other Nitrogen Supplements

Although cultured cells are normally capable of synthesizing all of the required amino acids, the addition of certain amino acids or amino acid mixtures may be used to further stimulate cell growth. The use of amino acids is particularly important for establishing cell cultures and protoplast cultures. Amino acids provide plant cells with an immediately available source of nitrogen, which generally can be taken up by the cells more rapidly than inorganic nitrogen (71).

The most common sources of organic nitrogen used in culture media are amino acid mixtures (e.g., casein hydrolysate), L-glutamine, L-asparagine, and adenine. Casein hydrolysate is generally used at concentrations between 0.05 and 0.1% (33). When amino acids are added alone, care must be taken as they can be inhibitory to cell growth. Examples of amino acids included in culture media to enhance cell growth are glycine at 2 mg/liter, glutamine up to 8 mM, asparagine at 100 mg/liter, L-arginine and cysteine at 10 mg/liter, and L-tyrosine at 100 mg/liter. Tyrosine has been used to stimulate morphogenesis in cell cultures but should only be used in an agar medium. Supplementation of the culture medium with adenine sulfate can stimulate cell growth and greatly enhance shoot formation (63).

Undefined Organic Supplements

Addition of a wide variety of organic extracts to culture media often results in favorable tissue responses. Supplements that have been tested include protein hydrolysates, coconut milk, yeast extracts, malt extracts, ground banana, orange juice, and tomato juice. However, undefined organic supplements should only be used as a last resort, and only coconut milk and protein hydrolysates are used to any extent today. Protein (casein) hydrolysates are generally added to culture media at a concentration of 0.05–0.1%, while coconut milk is commonly used at 5–20% (v/v).

The addition of activated charcoal (AC) to culture media may have either a beneficial or deleterious effect. Growth and differentiation have been stimulated in orchids (20, 79), onions (23, 24), carrots (23, 24), tomatoes (3), and ivy (14) but inhibited in tobacco (14), soybean (24), and Camellia (75) when AC was added to the culture medium. The effect of AC is generally attributed to one of three factors: absorption of inhibitory compounds (20, 23, 24, 40, 79), absorption of growth regulators from the culture medium (14, 80), or darkening of the medium (57). The inhibition of growth in the presence of AC is generally attributed to the absorption of phytohormones to AC. 1-Naphthaleneacetic acid (NAA), kinetin, 6-benzylaminopurine (BA), indole-3-acetic acid (IAA), and 6-γ-γ-dimethylallylamino purine (2iP) all bind to AC, with the latter two growth regulators binding quite rapidly (14, 80). The stimulation of cell growth by AC is generally attributed to its ability to bind to toxic phenolic compounds produced during culture. Activated charcoal is generally acid-washed and neutralized prior to addition to the culture medium at a concentration of 0.5–3.0%.

Solidifying Agents or Support Systems

Agar is the most commonly used gelling agent for preparing semisolid and solid plant tissue culture media. Agar has several advantages over other gelling agents. First, when agar is mixed with water, it forms a gel that melts at approximately 60°–100°C and solidifies at approximately 45°C; thus, agar gels are stable at all feasible incubation temperatures. Additionally, agar gels do not react with media constituents and are not digested by plant enzymes. The firmness of an agar gel is controlled by the concentration and brand of agar used in the culture medium and the pH of the medium. The agar concentrations commonly used in plant cell culture media range between 0.5 and 1.0%; these concentrations give a firm gel at the pHs typical of plant cell culture media. Horner et al. (38), however, noted that inclusion of 0.8–1.0% AC may reduce the firmness of the gel.

The purity of the agar used in culture media is of major importance. It has been demonstrated that agar contains Ca, Mg, K, and Na and that changing the agar concentration may effect nutrient availability. Others have suggested that agar contains carbohydrates and traces of amino acids and vitamins. Therefore, only highly purified agar should be used for critical experiments on tissue metabolism (16). Some impurities are removed by washing agar in double-distilled water for at least 24 hr, then rinsing in ethanol and drying at 60°C for 24 hr.

Gelatin at a concentration of 10% has been tried as a gelling agent for plant tissue culture media although it is not very useful because it melts at 25°C. Methocel and alginate also have been evaluated as gelling agents for culture media. Both compounds successfully supported the growth of plant cells, but there were problems in handling the methocel solutions (1). FMC Corp. has recently developed a highly purified agarose called Sea Plaque®, which has worked well in systems involving recovery of protoplasts when used at concentrations of 0.35– 0.7% (22). Another gelling agent commonly used for commercial as well as research purposes is Phytagel (Sigma Chemical) or Gelrite (Kelco Corp.). These products are synthetic and should be used at 1.25– 2.5 g/liter, resulting in a clear gel which aids in detecting contamination.

Alternative methods of support have included use of perforated cellophane (83), filter paper bridges (31), filter paper wicks (15), polyurethane foam (47), and polyester fleece (12, 13). Whether explants grow best on agar or on other supporting agents varies from one species of plant to the next.

Growth Regulators

Four broad classes of growth regulators are important in plant tissue culture: the auxins, cytokinins, gibberellins, and abscisic acid. Skoog and Miller (62) were the first to report that the ratio of auxin to cytokinin determined the type and extent of organogenesis in plant cell cultures. Both an auxin and a cytokinin are usually added to culture media in order to obtain morphogenesis, although the ratio of hormones required for root and shoot induction is not universally the same. Considerable variability exists among genera, species, and even cultivars in the type and amount of auxin and cytokinin required for induction of morphogenesis.

The auxins commonly used in plant tissue culture media are 1H-indole-3-acetic acid (IAA), 1H-indole-3-butyric acid (IBA), (2,4-dichlorophenoxy)acetic acid (2,4-D), and 1-naphthaleneacetic acid (NAA). The structures of these auxins are given in Fig. 2.1. The only naturally occurring auxin found in plant tissues is IAA. The other auxinlike compounds shown in Fig. 2.1 are synthetic but exhibit varying degrees of auxinlike activity. Other synthetic auxins that have been used in plant cell culture include 4-chlorophenoxyacetic acid or p-chlorophenoxyacetic acid (4-CPA, PCPA), (2,4,5,-trichlorophenoxy)acetic acid (2,4,5-T), 3,6-dichloro-2-methoxybenzoic acid (dicamba), and 4-amino-3,5,6-trichloropicolinic acid (picloram). The structures of these compounds are shown in Fig. 2.2.

The various auxins differ in their physiological activity and in the

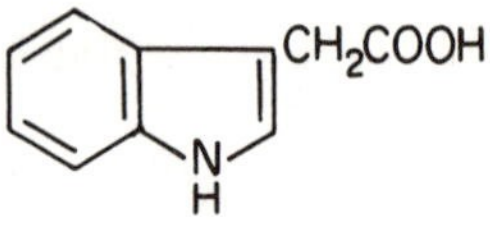

IAA MW 175.19

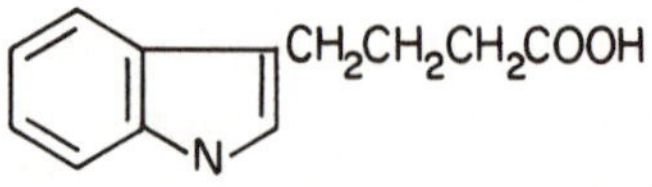

IBA MW 203.24

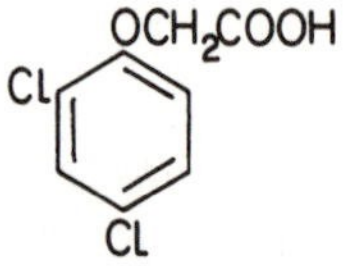

2,4-D MW 221.04

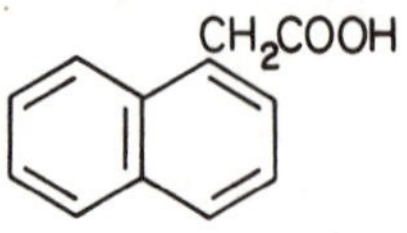

NAA MW 186.21

Fig. 2.1. Auxins that are commonly used in plant tissue culture.

extent to which they move through tissue, are bound to the cells, or metabolized. Based on stem curvature assays, 2,4-D has eight to twelve times the activity of IAA, 2,4,5-T has four times the activity, PCPA and picloram have two to four times the activity, and NAA has two times the activity of IAA (41). Although 2,4-D, 2,4,5-T, PCPA, and picloram are often used to induce rapid cell proliferation (37), exposure to high levels of or prolonged exposure to these auxins, particularly 2,4-D, results in suppressed morphogenetic activity (19, 32, 37, 64, 69). Auxins are generally included in a culture medium to stimulate callus

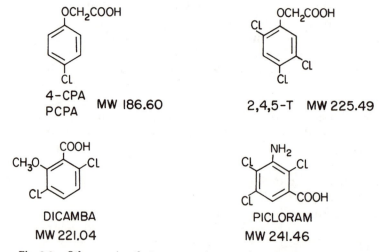

4-CPA
PCPA   MW 186.60

2,4,5-T   MW 225.49

DICAMBA
MW 221.04

PICLORAM
MW 241.46

Fig. 2.2. Other auxins that are sometimes used in plant tissue culture.

production and cell growth, to initiate shoots and particularly roots, and to induce somatic embryogenesis and stimulate growth from shoot apices and shoot tip cultures.

The cytokinins commonly used in culture media include 6-benzylamino purine or 6-benzyladenine (BAP, BA), 6-γ-γ-dimethylaminopurine (2iP), N-(2-furanylmethyl)-1H-purine-6-amine (kinetin), and 6-(4-hydroxy-3-methyl-*trans*-2-butenylamino)purine (zeatin). Structural formulas of these compounds are shown in Fig. 2.3, as are several names for 2iP commonly found in the literature. Zeatin and 2iP are considered to be naturally occurring cytokinins, while BA and kinetin

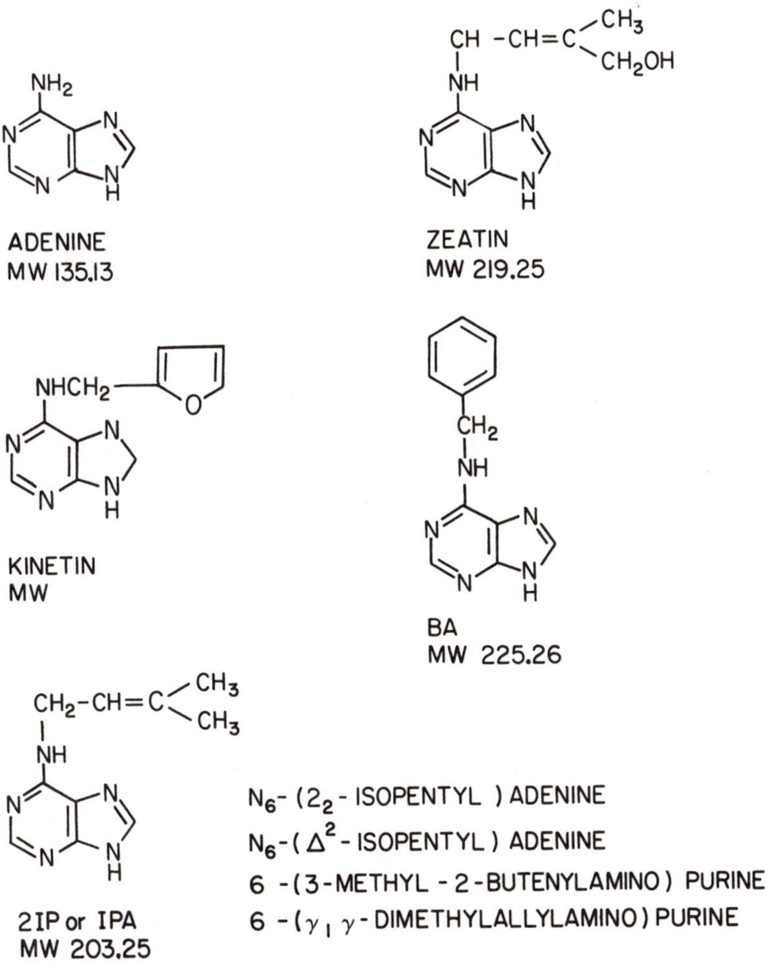

ADENINE
MW 135.13

ZEATIN
MW 219.25

KINETIN
MW

BA
MW 225.26

2IP or IPA
MW 203.25

N_6- (2_2 - ISOPENTYL) ADENINE

N_6-(Δ^2 - ISOPENTYL) ADENINE

6 -(3-METHYL - 2 -BUTENYLAMINO) PURINE

6 -(γ, γ - DIMETHYLALLYLAMINO) PURINE

Fig. 2.3. Cytokinins that are used in plant tissue culture.

are synthetically derived cytokinins. Adenine, another naturally occurring compound, has a base structure similar to that of the cytokinins and has shown cytokininlike activity in some cases. The side chain of the zeatin molecule contains a double bond and hence can exist in two isomeric forms (cis and trans). The trans-isomer of zeatin has the high biological activity. Some synthetic preparations of zeatin contain both the cis- and trans-isomers; but if these are utilized in cell culture media, a greater concentration is necessary compared with the concentration required if only *trans*-zeatin is used. Many plant tissues have an absolute requirement for a specific cytokinin for morphogenesis to occur, whereas some tissues are considered to be cytokinin independent, i.e., no cytokinin or a specific cytokinin may be required for organogenesis.

The cytokinins are generally added to a culture medium to stimulate cell division, to induce shoot formation and axillary shoot proliferation, and to inhibit root formation. The mechanism of cytokinin action is uncertain, although some compounds with cytokininlike activity have been found to be present in transfer-RNA (t-RNA). It is unclear, however, whether incorporation into transfer-RNA is required before specific cytokinin effects become apparent. Cytokinins have also been shown to activate RNA synthesis and to stimulate protein and enzyme activity in certain tissues.

The type of morphogenesis that occurs in a plant tissue culture largely depends upon the ratio and concentrations of auxins and cytokinins present in the medium. Root initiation of plantlets, embryogenesis, and callus initiation all generally occur when the ratio of auxin to cytokinin is high, whereas adventitious and axillary shoot proliferation occur when the ratio is low. The concentrations of auxins and cytokinins are equally as important as their ratio. The use of 2,4-D and BA both at 5.0 mg/liter promotes callus formation in *Agrostis*; however, the same growth regulators both used at 0.1 mg/liter promote shoot formation in the same tissue. In both cases, the ratio of auxin to cytokinin is one, but the concentration of auxin in the first case is so high that callus formation is favored regardless of the cytokinin concentration.

Gibberellins (GA$_3$) and abscisic acid (ABA) are two other growth regulators occasionally used in culture media. Plant tissue cultures can usually be induced to grow without either GA$_3$ or ABA, although certain species may require these hormones for enhanced growth. Generally, GA$_3$ is added to culture media to promote the growth of low-density cell cultures (70), to enhance callus growth (49, 51), and to elongate dwarfed or stunted plantlets (75). Abscisic acid is generally added to culture media to either inhibit or stimulate callus growth depending upon the species (2, 3, 39, 59), to enhance shoot or bud proliferation (34), and to inhibit latter stages of embryo development (25, 73).

MEDIA PREPARATION

The following plant tissue culture media are available as premixed powdered formulations: MS (53), B-5 (26), Nitsch and Nitsch's (55), White's (84), Heller's (36), Vacin and Went's (76), and Anderson's (4). Prepackaged media are quite convenient and valuable for specific procedures. Prepackaged salts and media offer not only the convenience of premeasured and mixed media but also the batch-to-batch consistency that cannot be obtained by making the media from stock solutions. Choose a prepackaged media supplier who follows GMP (good manufacturing procedures) in preparing the media; who quality control tests the media for elemental ion content, pH, osmolality, powder appearance and solubility; and who biologically tests the media with plant cell cultures.

The preparation of culture media requires clean glassware, high-quality water, pure chemicals, and careful measurement of all media components. The use of clean glassware is of utmost importance; the techniques for cleaning glassware were discussed in Chapter 1. Ultrapure double-distilled water should be used in preparing media for research purposes, but deionized water may be acceptable for large commercial operations. Tap water varies greatly in terms of dissolved impurities and particulate matter and should not be used for making culture media for any purpose. (See "Media Preparation Area" in Chapter 1 for discussion of water quality.) The chemicals used in preparing culture media should be analytical grade except for sucrose; refined grocery sugar is a sufficiently pure source of sucrose.

Culture media must contain the known macro- and micronutrients required for plant growth as well as a carbohydrate source, vitamins, agar (for semisolid media), and any hormones or plant extracts needed. The most widely used culture medium is Murashige and Skoog's MS medium (1), which was developed for tobacco culture but works well with most species. In the following sections, general procedures for preparing stock solutions of media components and for sterilizing media are presented. The specific formulations for several common media are given in the final sections of the chapter. Appendix 4 contains some general instructions for preparing media.

Stock Solutions

The use of stock solutions reduces the number of repetitive operations involved in media preparation and, hence, the chance of human or experimental error. Moreover, direct weighing of media components (e.g., micronutrients and hormones) that are required only in milligram or microgram quantities in the final formulation cannot be performed with sufficient accuracy for tissue culture work. For these components, preparation of concentrated stock solutions and subsequent dilution into the final media is standard procedure. In addition, concentrated solutions of some materials are more stable and can be stored for longer periods than more dilute solutions.

To prepare a stock solution, weigh out the required amount of the compound and place it in a clean volumetric flask. It is common practice to make a stock solution 10× or 100×, depending upon the solubility of the compound. Once the chemical is in the volumetric flask, dissolve it in a small amount of water, ethyl alcohol, 1 N NaOH, or 1 N HCl. Next, slowly add double-distilled water to the flask, while agitating. Continue this until the proper volume is reached. Label the flask with the name of the solution, preparation and expiration dates, and the name of the person who prepared the solution. Certain items, e.g., IAA, must be prepared and stored in amber bottles to prevent photodecomposition. The volumes of stock solutions prepared at various concentrations that must be used to achieve various final concentrations are presented in tabular form in Appendix 5.

Macronutrients

Stock solutions of macronutrients can be prepared at 10 times the concentration of the final medium. A separate stock solution for calcium salts may be required to prevent precipitation (28). Stock solutions of macronutrients can be stored safely for several weeks in a dark, cool place but are best stored in a refrigerator at 2°–4°C. Street (68) recommends filtration of all macronutrient stock solutions to remove any undissolved large particles before they are stored.

Micronutrients

Micronutrient stock solutions are generally made up at 100 times their final strength. It is recommended that micronutrient stocks be stored in either a refrigerator or freezer until needed. Mellor and Stace-Smith (50) found that micronutrient stock solutions could be stored in a refrigerator for up to 1 year without appreciable deterioration. Gamborg and Shyluk (28) recommend that stock solutions of potassium iodide be prepared, at a 100× concentration, and stored separately from other micronutrients. Iron stock solutions also should be prepared and stored separately from other micronutrients in an amber storage bottle. Perhaps the simplest method of including iron in a culture medium is to weigh out the desired amount of either sodium or potassium ferric EDTA and add it directly to the medium in the preparation process. Formulations for preparing stock solutions of iron are presented later.

Vitamins

Vitamins are prepared as 100× or 1000× stock solutions and stored in a freezer (−20°C) until used. Vitamin stock solutions should be made up each time media is prepared if a refrigerator or freezer is not available. Vitamin stock solutions can be stored safely in a refrigerator for 2–3 months but should be discarded after that time (28, 77). The recommended conditions for storing in dry form the vitamins commonly used in culture media are given in Appendix 6.

Growth Regulators

The auxins NAA and 2,4-D are considered to be stable and can be stored at 4°C for several months (29); IAA should be desiccated during storage at −20°C. Auxin stock solutions are generally prepared at 100–1000 times the final desired concentrations. Solutions of NAA and 2,4-D can be stored for several months in a refrigerator or indefinitely at −20°C. It is best to prepare fresh IAA solutions each time a medium is prepared; however, IAA solutions can be stored in an amber bottle at 4°C for no longer than a week. Generally, IAA and 2,4-D are dissolved in a small volume of 95% ethyl alcohol and then brought to volume with double-distilled water; NAA can be dissolved in a small amount of 1 N NaOH, which also can be used to dissolve 2,4-D and IAA. When using HCl or NaOH to dissolve auxins or cytokinins, the pH of the solution should be readjusted to 5.5–5.8 before making it up to volume or refrigerating the solution.

The cytokinins are considered to be stable and can be stored, desiccated, at −20°C. Cytokinin stock solutions are generally prepared at 100× to 1000× concentrations. Many of the cytokinins are difficult to dissolve, and a few drops of either 1 N HCl or 1 N NaOH are required to bring them into solution. Schmitz and Skoog (61) found that cytokinins can also be dissolved with a small amount of dimethylsulfoxide (DMSO) without injury to the plant tissue. An advantage of using DMSO is that it acts as a sterilant; thus, stock solutions containing DMSO can be added directly to the culture medium after sterilization has taken place.

Storage of Stock Solutions

Storage conditions for most stock solutions have already been pointed out; however, some additional points can be made. For convenience, many labs prepare stock solutions and then divide them into aliquots sufficient to prepare from 1 to 10 liters of medium; these aliquots are stored in small vials or plastic bags in a freezer. This procedure removes the inconvenience of having to unthaw a large volume of frozen stock each time medium is prepared. Gamborg and coworkers (27, 28) have recommended that 10× concentrated B-5 or MS medium, including all macro- and micronutrients, FeEDTA, vitamins, and sucrose, be prepared and stored at −20°C in plastic bags in 100-ml or 400-ml aliquots. To prepare 1 liter of media, 100 ml of the concentrate would be thawed, the necessary growth regulators added, agar added if required, the pH adjusted, and the medium brought up to volume with double-distilled water. Some have found that heating in a microwave oven is a satisfactory and quick method of thawing concentrated medium.

Sterilization of Media

Plant tissue culture media generally are sterilized by autoclaving at 121°C and 1.05 kg/cm² (15–20 psi). The time required for sterilization depends upon the volume of the medium in the vessel. The minimum times required for sterilization of different volumes of media are listed in Table 2.1. It is advisable to dispense media in small aliquots whenever possible because many media ingredients are broken down with prolonged exposure to heat and pressure. There is evidence that culture media exposed to extremely high temperatures does not gel (74) and that growth of cultured cells is considerably reduced in media autoclaved at temperatures in excess of 121°C (78). Lundergan and Wood (46) found that media could be sterilized by heating in a microwave oven for 2–4 min. However, the authors noted that heating media for less than 2 min resulted in contamination.

Several media components are considered thermolabile and should not be autoclaved. Thiamin has been reported to be heat labile and may break down rapidly if the pH of the culture medium is greater than 5.5 (17). Other studies, however, have found no reduction in growth when autoclaved thiamin was used in the culture medium (74). Other components considered thermolabile are fructose, calcium pantothenate, IAA, GA_3, riboflavin, folic acid, nicotinic acid, urea, L-glutamine, L-aspara-

Table 2.1. Minimum Sterilization Time for Plant Tissue Culture Media[a]

Volume of Media Per Vessel (ml)	Min. Sterilization Time[b] (min)
20–50	15
75	20
250–500	25
1000	30
1500	35
2000	40

[a] Biondi, S. and T. A. Thorpe. 1981. Requirements for a tissue culture facility. pp. 1–20. In Plant Tissue Culture: Methods and Applications in Agriculture. J. A. Thorpe (editor). Academic Press, New York.
[b] 121°C; 1.05 kg/cm² (21 psi).

gine, and adenine sulfate (43). Thermolabile media ingredients should be filter-sterilized rather than autoclaved. Stock solutions of the heat-labile components are prepared, then passed through a 0.45- or 0.22-μm filter into a sterile container (see Appendices 8 and 17). The filtered solution is then added with a sterile pipette to the culture medium, which has been autoclaved and allowed to cool to approximately 45°–50°C. The medium is then dispensed under sterile conditions.

Preparation of MS Medium According to Gamborg (28, 29)

Stock Solutions

a. Calcium chloride (g/liter):

$Ca_2Cl_2 \cdot 2H_2O$	150

b. Micronutrients, 100× (mg/100 ml):

Manganese sulfate	$MnSO_4 \cdot 4H_2O$	2230
Zinc sulfate	$ZnSO_4 \cdot 7H_2O$	860
Boric acid	H_3BO_3	620
Sodium molybdate	$Na_2MoO_4 \cdot 2H_2O$	25
Copper sulfate	$CuSO_4 \cdot 5H_2O$	2.5
Cobalt chloride	$CoCl_2 \cdot 6H_2O$	2.5

c. Potassium iodide (mg/100 ml):

KI	83

d. Vitamins, 100× (mg/100 ml):

Glycine	200
Nicotinic acid	50
Pyridoxine-HCl	50
Thiamin-HCl	10
Myo-inositol	10,000

Preparation of 1 Liter of Medium

1. Weigh out the following items and add directly to a 2-liter flask:

Potassium nitrate	KNO_3	1900 mg
Ammonium nitrate	NH_4NO_3	1650 mg
Magnesium sulfate	$MgSO_4 \cdot 2H_2O$	370 mg
Potassium phosphate	KH_2PO_4	170 mg
Sodium salt of ferrous EDTA	$Na_2FeEDTA \cdot 2H_2O$	43 mg

2. Dissolve the above in 400 ml of double-distilled water.
3. Weigh out and add 30 g sucrose to flask.

4. Add 2.9 ml of calcium chloride stock solution (a).
5. Add 1.0 ml of micronutrient stock solution (b).
6. Add 1.0 ml of potassium iodide stock solution (c).
7. Add 1.0 ml of vitamin stock solution (d).
8. Add desired volume of cytokinin (see Appendix 5).
9. Add desired volume of auxin (see Appendix 5).
10. Adjust volume to 800 ml with double-distilled water.
11. Adjust pH to 5.7 ± 0.1 with 1.0 N NaOH or 1.0 N HCl.
12. Adjust volume to 1 liter with double-distilled water.
13. Add 8–10 g agar if required.
14. Cool and dispense into culture tubes.
15. Autoclave.

Preparation of MS Medium According to Dodds and Roberts (17)

Stock Solutions

a. Micronutrients, 100× (mg/liter):

Manganese sulfate	$MnSO_4 \cdot 4H_2O$	2230
Zinc sulfate	$ZnSO_4 \cdot 4H_2O$	860
Boric acid	H_3BO_3	620
Potassium iodide	KI	83
Sodium molybdate	$Na_2MoO_4 \cdot 2H_2O$	25
Copper sulfate	$CuSO_4 \cdot 5H_2O$	2.5
Cobalt chloride	$CoCl_2 \cdot 6H_2O$	2.5

b. Iron/EDTA, 200× (mg/100 ml):

Iron sulfate	$FeSO_4 \cdot 7H_2O$	557
Sodium EDTA	$Na_2EDTA \cdot 2H_2O$	745

c. Vitamins, 100× (mg/100 ml):

Glycine	200
Nicotinic acid	50
Pyridoxine-HCl	50
Thiamin-HCl	10

Preparation of 1 Liter of Medium

1. Weigh out the following items and add directly to a 2-liter flask:

Potassium nitrate	KNO_3	1900 mg
Ammonium nitrate	NH_4NO_3	1650 mg
Calcium chloride	$CaCl_2 \cdot 2H_2O$	440 mg
Magnesium sulfate	$MgSO_4 \cdot 7H_2O$	370 mg
Potassium phosphate	KH_2PO_4	170 mg

2. Dissolve the above in approximately 400 ml of double-distilled water.
3. Add 10 ml of micronutrient stock solution (a).
4. Add 5 ml of iron/EDTA stock solution (b).
5. Add cytokinins as required (see Appendix 5).
6. Weigh out desired amount of auxin, dissolve, and add directly to medium.
7. Adjust volume to 800 ml with double-distilled water.
8. Adjust pH to 5.7 ± 0.1 with 1.0 N NaOH or 1.0 N HCl.
9. Bring volume to 1 liter with double-distilled water and store in refrigerator if necessary.
10. Divide into 10 aliquots of 100 ml each and place each in a 250-ml flask.
11. Add 3 g of sucrose and 0.8 g of agar (if required) to each 250-ml flask.
12. Autoclave.
13. Cool medium to 50°C; add 1 ml of filter-sterilized vitamin stock solution (c) to each 250-ml flask.
14. Dispense into sterile culture tubes under aseptic conditions.

Preparation of MS Medium According to Helgeson (35) (Prepares a 0.5× MS medium)

Stock Solutions

a. Ammonium nitrate (g/liter):

NH_4NO_3 82.5

b. Potassium nitrate (g/liter):

KNO_3 95

c. Calcium chloride (g/liter):

$Ca_2Cl_2 \cdot 2H_2O$ 88

d. Potassium phosphate (g/liter):

KH_2PO_4 34

e. Micronutrients (g/liter):

Boric acid	H_3BO_3	1.24
Sodium molybdate	$Na_2MoO_4 \cdot 2H_2O$	0.05
Colbalt chloride	$CoCl_2 \cdot 6H_2O$	0.005
Potassium iodide	KI	0.166

f.   Micronutrients (g/liter):

Manganese sulfate	$MnSO_4 \cdot H_2O$	3.38
Magnesium sulfate	$MgSO_4 \cdot 7H_2O$	74.0
Copper sulfate	$CuSO_4 \cdot 5H_2O$	0.005
Zinc sulfate	$ZnSO_4 \cdot 7H_2O$	1.725

g.   Iron/EDTA (g/liter):

Sodium EDTA	$Na_2EDTA \cdot 2H_2O$	1.865
Iron sulfate	$FeSO_4 \cdot 7H_2O$	1.39

Preparation of 1 Liter of Medium

1.   Pour 500 ml of double-distilled water into a 2-liter flask and begin stirring.
2.   Weigh out 30 g of sucrose; add to flask and dissolve.
3.   Weigh out and dissolve the following items, then add to flask:

Myo-inositol	100.0 mg
Thiamin-HCl	0.1 mg
Pyridoxine-HCl	0.5 mg
Glycine	2.0 mg
Nicotinic acid	0.5 mg

4.   Add 20 ml each of stock solutions a, b, and g.
5.   Add 5 ml each of stock solutions c, d, e, and f.
6.   Weigh out, dissolve, and add any auxins or cytokinins required.
7.   Bring volume to 900 ml with double-distilled water.
8.   Adjust pH to 5.8.
9.   Pour medium into a graduated cylinder and adjust volume to 1 liter.
10.   Pour medium into a 2-liter Erlenmeyer flask, add 10 g of agar if required, and steam in autoclave at 100°C for 25 min.
11.   Cool to 90°C and (1) dispense into culture tubes then autoclave or (2) autoclave, cool to 50°C, add any heat-labile components, and dispense.

Preparation of MS Medium

Stock Solutions

a.   Macronutrients, 10× (g/liter):

Potassium nitrate	KNO_3	19.0
Ammonium nitrate	NH_4NO_3	16.5
Calcium chloride	$CaCl_2 \cdot 2H_2O$	4.4
Magnesium sulfate	$MgSO_4 \cdot 7H_2O$	3.7

b. Phosphate, $10\times$ (g/liter):

| Potassium phosphate | KH_2PO_4 | 1.7 |
| Sodium phosphate | NaH_2PO_4 | 1.7 |

c. Micronutrients (g/liter):

Boric acid	H_3BO_3	0.62
Manganese sulfate	$MnSO_4 \cdot H_2O$	1.69
Zinc sulfate	$ZnSO_4 \cdot 7H_2O$	0.86

d. Micronutrients (g/liter):

| Potassium iodide | KI | 0.83 |
| Sodium molybdate | $Na_2MoO_4 \cdot 2H_2O$ | 0.25 |

e. Micronutrients (g/liter):

| Copper sulfate | $CuSO_4 \cdot 5H_2O$ | 0.25 |
| Cobalt chloride | $CoCl_2 \cdot 6H_2O$ | 0.25 |

f. Iron/EDTA (g/100 ml):

| Sodium EDTA | Na_2EDTA | 0.372 |
| Iron sulfate | $FeSO_4 \cdot 7H_2O$ | 0.278 |

To prepare the iron stock solution weigh out the desired amounts of sodium EDTA and iron sulfate separately, and dissolve each in 50 ml of double-distilled water. Once both solutions are completely dissolved, slowly combine the two with continuous stirring. Store in amber bottle at 4°C.

g. Vitamins, $1000\times$ (mg/100 ml):

Glycine	200.0
Nicotinic acid	50.0
Pyridoxine-HCl	50.0
Thiamin-HCl	10.0
Myo-inositol	10,000.0

Preparation of 1 Liter of Medium

1. Pour 400 ml of double-distilled water into a 2-liter Erlenmeyer flask and begin stirring.
2. Add 100 ml each of stock solutions a and b.
3. Add 10 ml each of stock solutions c and f.
4. Add 1 ml each of stock solutions d and g.
5. Add 0.1 ml of stock solution e.
6. Weigh out and add 30 g of sucrose; dissolve.
7. Bring volume to 900 ml with double-distilled water.

8. Add auxins and cytokinins as required (see Appendix 5).
9. Adjust pH to 5.7 ± 0.1.
10. Add 8.0 g of agar if required; bring to boil and then cool to 60°C.
11. Bring volume to 1 liter with double-distilled water.
12. Dispense into culture tubes and autoclave.

Modifications

Stock solutions a and b can be prepared at 100× concentration instead of 10× concentration. If this is done, add 10 ml (instead of 100 ml) in step 2. To avoid precipitation if a 100× concentrated stock solution a is used, omit the calcium chloride and prepare a separate calcium chloride stock.

Gamborg's B-5 Medium (26)

Stock Solutions

a. Micronutrients (mg/100 ml):

Manganese sulfate	$MnSO_4 \cdot H_2O$	1000
Boric acid	H_3BO_3	300
Zinc sulfate	$ZnSO_4 \cdot 7H_2O$	200
Sodium molybdate	$Na_2Mo_4 \cdot 2H_2O$	25
Copper sulfate	$CuSO_4 \cdot 5H_2O$	2.5

Store solution in freezer.

b. Vitamins (mg/100 ml):

Nicotinic acid	100
Thiamin-HCl	1000
Pyridoxine-HCl	100
Myo-inositol	10,000

Store solution in freezer.

c. Calcium chloride (g/100 ml):

$CaCl_2 \cdot 2H_2O$	15

d. Potassium iodide (mg/100 ml):

KI	75

Preparation of 1 Liter of Medium

1. Pour 400 ml of double-distilled water into a 2-liter Erlenmeyer flask and begin stirring.
2. Weigh out and add the following ingredients:

Sodium phosphate	$NaH_2PO_4 \cdot H_2O$	150 mg
Potassium nitrate	KNO_3	2500 mg

Ammonium sulfate	$(NH_4)_2SO_4$	134 mg
Magnesium sulfate	$MgSO_4 \cdot 7H_2O$	250 mg
Ferric EDTA		43 mg
Sucrose		20 g

3. Add 1 ml of calcium chloride stock solution (c).
4. Add 1 ml of micronutrient stock solution (a).
5. Add 1 ml each of potassium iodide (d) and vitamin stock solution (b).
6. Add auxins and cytokinins as desired (see Appendix 5).
7. Bring volume to 900 ml with double-distilled water.
8. Adjust pH to 5.5 with 0.2 N KOH or 0.2 N HCl.
9. Add 6–8 g of agar if required.
10. Bring volume to 1 liter with double-distilled water.
11. Dispense medium into culture vessels and autoclave.

Modifications

1. Add 1.0 g/liter of N-Z amine® Type A, a pancreatic hydrolyzate of casein.
2. In step 2, omit KNO_3 and $(NH_4)_2SO_4$ and use instead 20 mM KCl, 10 mM citric acid, and 20 mM NH_4OH.

Anderson's Rhododendron Medium (4)

Stock Solutions

a. Macronutrients, 100× (g/liter):

Ammonium nitrate	NH_4NO_3	40.0
Potassium nitrate	KNO_3	48.0
Magnesium sulfate	$MgSO_4 \cdot 7H_2O$	37.0
Sodium phosphate	$NaH_2PO_4 \cdot H_2O$	38.0

b. Calcium chloride, 100× (g/liter):

| $CaCl_2 \cdot 2H_2O$ | 44.0 |

c. Iron/EDTA, 100× (g/liter):

| Iron sulfate | $FeSO_4 \cdot 7H_2O$ | 5.57 |
| Sodium EDTA | Na_2EDTA | 7.45 |

d. Micronutrients (mg/100 ml)

Manganese sulfate	$MnSO_4 \cdot H_2O$	1690
Zinc sulfate	$ZnSO_4 \cdot 7H_2O$	860
Boric acid	H_3BO_3	620
Cobalt chloride	$CoCl_2 \cdot 6H_2O$	2.5

Copper sulfate	$CuSO_4 \cdot 5H_2O$	2.5
Sodium molybdate	$Na_2MoO_4 \cdot 2H_2O$	25

e. Potassium iodide (mg/100 ml):

KI	30

f. Vitamins (mg/100 ml):

Myo-inositol	10,000
Thiamin-HCl	40

Preparation of 1 Liter of Medium

1. Measure out 400 ml of double-distilled water and pour into a 2-liter Erlenmeyer flask and begin stirring.
2. Weigh out and add 30 g of sucrose and stir until dissolved.
3. Add 10 ml each of stock solutions a, b, and c.
4. Add 1 ml each of stock solutions d and e.
5. Add 1 ml of vitamin stock solution (f).
6. Bring volume to 900 ml with double-distilled water.
7. Add auxins and cytokinins if desired (see Appendix 5).
8. Adjust pH to 4.5 with 1 N NaOH or 1 N HCl.
9. Add 6–10 g of agar if desired.
10. Bring volume to 1 liter with double-distilled water.
11. Dispense into culture tubes and autoclave.

Modifications

1. For Stage I micropropagation, add 2 mg/liter 2iP and 0.5 mg/liter IAA. Use at 1/2 strength without agar and dispense 5 ml of media/tube.
2. For Stage II (first subculture), use medium at full strength with 8–12 g/liter agar, plus 15 mg/liter 2iP and 4 mg/liter IAA.
3. For Stage II (subsequent cultures, multiplication stage), use medium at full strength with agar, plus 1–15 mg/liter 2iP and 0.5–4.0 mg/liter IAA.
4. For Stage III (pretransplant stage), prepare medium at 1/4 strength omitting sucrose and add 22.5 g/liter sucrose, 10 g/liter agar, and 600 mg/liter activated charcoal. Do not add any growth regulators.

Vacin and Went's Orchid Medium (76)

Stock Solutions

a. Macronutrients (g/liter):

Ammonium sulfate	$(NH_4)_2SO_4$	50.0
Magnesium sulfate	$MgSO_4 \cdot 7H_2O$	25.0

Potassium nitrate	KNO_3	52.5
Potassium phosphate	KH_2PO_4	25.0
Manganese sulfate	$MnSO_4 \cdot 4H_2O$	0.75

b. Calcium phosphate (g/liter):

$Ca_3(PO_4)_2$	20.0

c. Iron/EDTA (g/liter):

FeNaEDTA	3.7

Preparation of 1 Liter of Medium

1. Measure out 500 ml of double-distilled water and pour into a 2-liter Erlenmeyer flask and begin stirring.
2. Weigh out and add 20 g of sucrose and stir until dissolved.
3. Add 10 ml each of stock solutions a, b, and c.
4. Bring volume to 900 ml with double-distilled water.
5. Add 0.4 mg of thiamin and stir until dissolved.
6. Adjust pH to 5.8.
7. Add 6–10 g agar if required.
8. Add auxins or cytokinins if required.
9. Bring volume to 1 liter with double-distilled water.
10. Dispense medium into culture tubes and autoclave.

Nitsch and Nitsch's Medium (55)

Stock Solutions

a. Macronutrients (g/liter):

Ammonium nitrate	NH_4NO_3	72.0
Magnesium sulfate	$MgSO_4 \cdot 7H_2O$	18.5
Potassium nitrate	KNO_3	95.0
Potassium phosphate	KH_2PO_4	6.8

b. Micronutrients (mg/100 ml):

Boric acid	H_3BO_3	1000
Copper sulfate	$CuSO_4 \cdot 5H_2O$	2.5
Manganese sulfate	$MnSO_4 \cdot H_2O$	2500
Sodium molybdate	$Na_2MoO_4 \cdot 2H_2O$	25
Zinc sulfate	$ZnSO_4 \cdot 7H_2O$	1000

c. Iron/EDTA (g/liter):

Iron sulfate	$FeSO_4 \cdot 7H_2O$	2.78
Sodium EDTA	Na_2EDTA	3.73

d.   Calcium chloride (g/liter):

$Ca_2Cl_2 \cdot 2H_2O$ 16.6

Preparation of 1 Liter of Medium

1.   Measure out 500 ml of double-distilled water and pour into a 2-liter Erlenmeyer flask and being stirring.
2.   Weigh out and add 30 g of sucrose and stir until dissolved.
3.   Add 10 ml of macronutrient stock solution (a).
4.   Add 1 ml of micronutrient stock solution (b).
5.   Add 10 ml each of stock solutions c and d.
6.   Add any vitamins if required.
7.   Add any auxins and cytokinins if required.
8.   Bring volume to 900 ml with double-distilled water.
9.   Adjust pH to 5.7 ± 0.1.
10. Add 6.0–10.0 g agar if required.
11. Bring volume to 1 liter with double-distilled water.
12. Dispense into culture tubes and autoclave.

REFERENCES

1.   Adaoha, Mbanaso E. N. and D. H. Roscoe. 1982. Phytochrome regulation of uptake of metabolites by coconut nuclei *in vitro*. Plant Sci. Lett. 25:61–66.
2.   Altman, A. and R. Goren. 1971. Promotion of callus formation by ABA in *Citrus* bud cultures. Physiol. Plant. 47:844–846.
3.   Anagnostakis, S. L. 1974. Haploid plants from anthers of tobacco—enhancement with charcoal. Planta 115:281.
4.   Anderson, W. C. 1975. Propagation of rhododendron: Part 1. Development of culture medium for multiplication of shoots. Proc. Int. Plant Prop. Soc. 25:129–134.
5.   Ball, E. 1953. Hydrolysis of sucrose by autoclaving media—a neglected aspect in the technique of culture of plant tissues. Bull. Torrey Bot. Club 80:409–411.
6.   Bergmann, L. 1967. Wachstum gruner Suspensionskulturen von *Nicotiana tabacum* var. "Samsun" mit CO_2 als Kohlenstoffquelle. Planta 74:243–249.
7.   Blumenfeld, A. and S. Gazit. 1970. Interaction of kinetin and abcisic acid in the growth of soybean callus. Plant Physiol. 45:535–536.
8.   Bonner, J. 1937. Vitamin B—a growth factor for higher plants. Science 85:183–184.
9.   Bonner, J. 1938. Thiamine (vitamin B_1) and the growth of roots: the relation of chemical structure to physiological activity. Am J. Bot. 25:543–549.
10. Bonner, J. 1940. On the growth factor requirements of isolated roots. Am. J. Bot. 27:692–701.
11. Chandler, M. T., N. Tandeau De Marsac, and Y. Dekouchkovsky. 1972. Photosynthetic growth of tobacco cells in liquid suspension. Can. J. Bot. 50:2265–2270.
12. Cheng, T. Y. 1978. Clonal propagation of woody plant species through tissue culture techniques. Comb. Proc. Int. Plant Prop. Soc. 28:139–155.
13. Cheng, T. Y. and T. H. Voqui. 1977. Regeneration of Douglas fir plantlets through tissue culture. Science 198:306–307.

14. Constantin, M. J., R. R. Henke, and M. A. Mansur. 1977. Effect of activated charcoal on callus growth and shoot organogenesis in tobacco. *In Vitro* 13:293.
15. Davis, M. J., R. Baker, and J. J. Hanan. 1977. Clonal multiplication of carnation by micropropagation. J. Am. Soc. Hort. Sci. 102:48–53.
16. Day, D. 1942. Thiamin content of agar. Bull. Torrey Bot. Club 69:11–20.
17. Dodds, J. H. and L. W. Roberts. 1982. Experiments in Plant Tissue Culture. Cambridge Univ. Press, Cambridge, London, New York.
18. Dougall, D. K. and D. C. Verma. 1978. Growth and embryo formation in wild-carrot suspension cultures with ammonium ion as the sole nitrogen source. *In Vitro* 14:180.
19. Engvild, K. C. 1978. Substituted indoleacetic acids tested in tissue cultures. Physiol. Plant. 44:345.
20. Ernst, R. 1974. The use of activated charcoal in a symbiotic seedling culture of *Paphropedilum*. Am. Orchid Soc. Bull. 43:35.
21. Ferguson, J. D., H. E. Street, and S. B. David. 1958. The carbohydrate nutrition of tomato roots. V. The promotion and inhibition of excised root growth by various sugars and sugar alcohols. Ann. Bot. 22:513–524.
22. FMC Technical Report. 1982. The Agarose Monograph. FMC Corporation, Bio Products Department, Rockland, ME.
23. Fridborg. G. 1978. Effects of activated charcoal on morphogenesis in plant tissue cultures. *In* Proc. Int. Congr. Plant Tissue and Cell Culture. University of Calgary, Canada.
24. Fridborg, G. and T. Erikson. 1975. Effects of activated charcoal on growth and morphogenesis in cell cultures. Physiol. Plant. 34:306.
25. Fujimura, T. and A. Komamine. 1975. Effects of various growth regulators on the embryogenesis in a carrot cell suspension culture. Plant Science Lett. 5:359–364.
26. Gamborg, O. L., R. A. Miller, and K. Ojima. 1968. Nutrient requirements of suspension cultures of soybean root cells. Exp. Cell Res. 50:151–158.
27. Gamborg, O. L., T. Murashige, T. A. Thorpe, and I. K. Vasil. 1976. Plant tissue culture media. *In Vitro* 12:473.
28. Gamborg, O. L. and J. P. Shyluk. 1981. Nutrition, media and characteristics of plant cell and tissues culture. pp. 21–44. *In* Plant Tissue Culture: Methods and Applications in Agriculture. by T. A. Thorpe (editor). Academic Press, New York.
29. Gamborg, O. L. and L. R. Witter. 1975. Plant Tissue Culture Methods. National Research Council of Canada, Saskatoon, Canada.
30. Gautheret, R. J. 1942. Manuel Technique de Culture des Tissue Vegetaux. Masson Cie, Paris.
31. Goodwin, P. B. 1976. An improved medium for the rapid growth of isolated potato buds. J. Exp. Bot. 17:590–595.
32. Halperin, W. 1966. Alternative morphogenetic effects in cell suspension cultures. Am. J. Bot. 53:443.
33. Haung, Li-Chun and T. Murashige. 1977. Plant tissue culture media; their preparation and some applications. Tissue Culture Assoc. Man. 3:539.
34. Heide, O. M. 1968. Stimulation of adventitious bud formation in begonia leaves by abscisic acid. Nature 219:960–961.
35. Helgeson, J. P. 1979. Tissue and cell-suspension culture. *In* Nicotiana: Procedures for Experimental Use. R. D. Durbin (editor). U.S. Dept. Agric., Washington, DC.
36. Heller, R. 1953. Recherches sur la nutrition minerale des tissus vegetaux cultivees *in vitro*. Am. Sci. Nat. Bot. Biol. Veg. 14:1–223.
37. Hill, G. P. 1967. Morphogenesis in stem-callus cultures of *Convolvulus arvensis* L. Ann. Bot. 31:437.
38. Horner, M., J. A. McComb, A. J. McComb, and H. E. Street. 1977. Ethylene produc-

tion and plantlet formation by Nicotiana anthers cultured in the presence and absence of charcoal. J. Exp. Bot. 28:1365–1372.

39. Isikawa, H. 1974. *In vitro* formation of adventitious buds and roots on the hypocotyl of *Cryptomeria japonica*. Bot. Mag. (Tokyo) 87:73–77.

40. Klein, B. and M. Bopp. 1971. Effect of activated charcoal in agar on the culture of lower plants. Nature (London) 230:474.

41. Lam, T. H. and H. E. Street. 1977. The effect of selected aryloxyalkane-carboxylic acids on the growth and levels of soluble phenols in cultured cells of *Rosa damescens*. Z. Pflanzenphysiol. 84:121.

42. Larosa, P. C., P. M. Hasegawa, and R. Bressan. 1981. Initiation of photoautotropic potato cells. HortScience 16:433.

43. Liau, D. F. and W. G. Boll. 1970. Callus and cell suspension culture of bushbean (*Phaseorlus vulgaris*). Can. J. Bot. 48:1119–1130.

44. Lloyd, G. and B. McCown. 1980. Use of microculture for production and improvement of *Rhododendron* spp. HortScience 15:416.

45. Lloyd, G. and B. McCown. 1981. Commercially feasible micropropagation of mountain laurel, *Kalmia latiflia*, by use of shoot tip cultures. Comb. Proc. Int. Plant Prop. Soc. 30:421–427.

46. Lundergan, C. and N. J. Wood. 1981. Microwave sterilization of tissue culture media. HortScience 16:417.

47. McComb, J. A. and S. Newton. 1981. Propagation of kangaroo paws using tissue culture. J. Hort. Sci. 56:181–183.

48. McCown, B. and G. Lloyd. 1981. Woody plant medium (WPM)—a mineral nutrients formulation for microculture of woody plant species. HortScience 16:453.

49. Mehra, A. and P. N. Mehra. 1972. Differentiation in callus cultures of *Mesembryanthemum floribundum*. Phytomorphology 22:171–176.

50. Mellor, F. C. and R. Stace-Smith. 1977. Virus-free potatoes by tissue culture. pp. 616–635. *In* Plant Cell Tissue And Organ Culture. J. Reinert and Y. P. S. Bajay (editors). Springer-Verlag, New York.

51. Murashige, T. 1964. Analysis of the inhibition of organ formation in tobacco tissue culture by gibberellin. Physiol. Plant. 17:636–643.

52. Murashige, T. 1974. Plant propagation through tissue culture. Ann. Rev. Plant Physiol. 25:135–166.

53. Murashige, T. and F. Skoog. 1962. A revised medium for rapid growth and bioassays with tobacco cultures. Physiol. Plant. 15:473.

54. Nitsch, J. P. 1969. Experimental androgenesis in *Nicotiana*. Phytomorphology 19:389.

55. Nitsch, J. P. and C. Nitsch. 1969. Haploid plants from pollen grains. Science 163:85–87.

56. Ohira, K., M. Ikeda, and K. Ojima. 1976. Thiamine requirements of various plant cells in suspension cultures. Plant Cell Physiol. 17:583.

57. Proskauer, J. and R. Berman. 1970. Agar culture medium modified to approximate soil conditions. Nature (London) 227:1161.

58. Robbins, W. J. and M. A. Bartley. 1937. Vitamin B and the growth of excised tomato roots. Science 85:246–247.

59. Sankhla, N. and D. Sankhla. 1968. Naturwissenschaften 55:91–92.

60. Schenk, R. V. and A. C. Hilderbrandt. 1972. Medium and techniques for induction and growth of monocotyledonous and dicotyledonous plant cell cultures. Can. J. Bot. 50:199–204.

61. Schmitz, R. Y. and F. Skoog. 1970. The use of dimethyl sulfoxide as a solvent in the tobacco bioassay for cytokinins. Plant Physiol. 45:537–538.

62. Skoog, F. and C. O. Miller. 1957. Chemical regulation of growth and organ formation in plant tissue grown *in vitro*. Symp. Soc. Exp. Biol. 11:118.
63. Skoog, F. and C. Tasi. 1948. Chemical control of growth and bud formation in tobacco stem segments and callus cultured *in vitro*. Am. J. Bot. 35:782.
64. Staba, E. J., P. Laursen, and S. Buchner. 1965. Medicinal plant tissue cultures. p. 191. *In* Int. Conf. on Plant Tissue Culture. P. R. White and A. R. Grove (editors). McCutcham Press, Berkeley, CA.
65. Steiner, A. A. and H. van Winden. 1970. Recipe for ferric salts of ethylene diaminetetraacetic acid. Plant Physiol. 46:862.
66. Steward, F. C. 1969. Plant Physiology. Vol. VB. Academic Press, New York.
67. Stoltz, L. P. 1979. Iron nutrition of *Cattleya* orchids grown *in vitro*. J. Am. Soc. Hort. Sci. 104:308.
68. Street, H. E. 1977. Plant Tissue and Cell Culture. Botanical Monographs. Vol. II. 2nd ed. Blackwell Publications, Oxford.
69. Street, H. E. 1979. Embryogenesis and chemically induced organogenesis. p. 123. *In* Plant Cell and Tissue Culture: Principles and Applications. W. R. Sharp, P. O. Larsen, E. F. Paddock, and V. Raghaven (editors). Ohio State Univ. Press, Columbus.
70. Stuart, R. and H. E. Street. 1971. Studies on the growth in culture of plant cells. X. Further studies on the conditioning of culture media by suspensions of *Acer pseudoplantanus* L. J. Exp. Bot. 20:556–571.
71. Thom, M., A. Maretzki, E. Komer and W. S. Sakai. 1981. Nutrient uptake and accumulation by sugarcane cell cultures in relation to growth cycle. Plant Cell Tissue and Organ Culture 1:3–14.
72. Thorpe, T. A. 1978. Physiological and biochemical aspects of organogenesis *in vitro*. p. 49. *In* Frontiers of Plant Tissue Culture; Proc. 4th Int. Congr. Plant Tissue and Cell Culture. T. A. Thorpe (editor). University of Calgary, Canada.
73. Tisserat, B. and T. Murashige. 1977. Effects of ethephon and 2,4-dichlorophenoxyacetic acid on asexual embryogenesis *in vitro*. Plant Physiol. 60:437–439.
74. Torres, K. C. 1985. Unpublished results.
75. Torres, K. C. and J. A. Carlisi. 1986. Enhanced shoot multiplication and rooting of *Camellia sasanqua*. Plant Cell Reports 5:381–384.
76. Vacin, E. F. and F. W. Went. 1949. Some pH changes in nutrient solutions. Bot. Gaz. 110:605–613.
77. Vasil, I. K. and A. C. Hilderbrandt. 1966. Variations of morphogenetic behavior in plant tissue cultures. I. *Cichorium endiva*. Am. J. Bot. 53:860–869.
78. Veliky, I. A. and S. M. Martin. 1970. A fermenter for plant cell suspension cultures. Can. J. Microbiol. 16:223–226.
79. Wang, P. J. and L. C. Huang. 1976. Beneficial effects of activated charcoal on plant tissue and organ cultures. *In Vitro* 12:260.
80. Weatherhead, M. A., J. Burdon, and G. G. Menshaw. 1978. Some effects of activated charcoal as an additive to plant tissue culture media. Z. Pflanzenphysiol. 89:141.
81. White, P. R. 1937. Vitamin B_1 in the nutrition of excised tomato roots. Plant Physiol. 12:793–802.
82. White, P. R. 1943. A Handbook of Tissue Culture. Ronald Press, New York.
83. White, P. R. 1953. A comparison of certain procedures for the maintenance of plant tissue cultures. Am. J. Bot. 40:517–524.
84. White, P. R. (editor). 1963. The Cultivation of Animal and Plant Cells. 2nd ed. Ronald Press, New York.

3

Stages of Micropropagation

The development of plantlets *in vitro* can be divided into three major steps or stages (28). In Stage I, called the explant or establishment stage, a suitable plant part (e.g., explant) is disinfected and cultured aseptically in a culture medium. Stage I material is then utilized for Stage II, which is termed the multiplication phase. The objective of Stage II is to rapidly increase the number of propagules by somatic cell embryogenesis, enhanced axillary branching, or adventitious bud formation. Stage II material may be recycled by subculturing the material back on a proliferation medium or it may be passed to Stage III. Stage III is called the conditioning or pretransplant stage; in some cases special dormancy conditions may have to be satisfied before rooting will occur. Recently, a fourth stage, Stage IV, has been described during which acclimatization of the plantlet to *in vivo* conditions occurs. In some cases, *in vivo* rooting may occur during Stage IV.

The stages of micropropagation were first described by Murashige (28) in 1974. He developed three stages of micropropagation for the *in vitro* multiplication of plants. These stages have now been adopted for a wide variety of plant material propagated both at the research and commercial level. The stages as described by Murashige not only describe the procedural steps of micropropagation, but these steps generally coincide with points in the process where environmental or media changes are required. The three stages described by Murashige include the establishment stage (Stage I) whereby the explant is established under aseptic conditions. Stage II or the multiplication stage is where the number of propagules is increased or multiplied. Stage III or the rooting stage is where the *in vitro*-derived shoots are rooted and conditioning of the plantlets for transfer out of the culture tube begins. Two other stages can now be added to the process: Stage 0 which involves the preparation of the stock or mother plant from which the primary explant is to be derived and Stage IV which is a stage that involves the transfer of the plantlets to an environment external of the culture tube.

STAGE 0—STOCK PLANT SELECTION AND PREPARATION

Careful attention should be made to make certain the stock plant is a typical variety or cultivar, clearly resembling other plants of the same species or cultivar. The stock plant should be disease-free, preferably maintained in either a growth chamber or greenhouse. The stock plants should be on a regularly maintained pesticide and fertilizer program.

The time of the year in which the explant is taken may effect the results of the micropropagation program. Changes in temperature, day-length, light intensity, and water availability throughout a year will affect the levels of carbohydrates, proteins, and growth substances in the stock plant thus subsequently affecting the response of the explant *in vitro*. Best results are generally achieved when the explant is taken during the active phase of growth. A possible exception is when the explant is to be derived from a storage organ. Explants can be derived from plant tissue which is in the dormant phase of growth but certain points must be considered. If the material is taken during the dormant phase of growth, the dormancy requirements may have to be met or broken. Plant material taken during the dormant or resting phase may be broken by removing bud scales which may contain bud break inhibitors. Soaking shoot tips in GA_3 or placing the shoot in a refrigerator are techniques which may also be used to break dormancy.

STAGE I—ESTABLISHMENT OF AN ASEPTIC CULTURE

The first step in any successful tissue culture program is the selection of a suitable explant source. Almost any plant tissue or organ can be used as an explant, but the degree of success obtained will depend upon the culture system used, the species being cultured, and the removal of surface contaminants from the explant. The primary goal of Stage I is to obtain a large percentage of explants free from surface pathogens. Disinfecting the surface generally involves washing the tissue, followed by sterilization with one or more disinfectants.

Washing the explant under running tap water for 30 min to 2 hr greatly reduces the amount of contamination on explants derived from field-grown material, highly pubescent tissue, roots, and/or storage organs. This technique has been successful in reducing contamination in the Gesneriaceae and the *Lilium* genus (15). Washing the explant with soapy water before placing it under running tap water may further reduce the number of pathogens present on the explant or make them more accessible to sterilant. Other presterilization techniques that have been used by the author include soaking the tissue in a mild fungicide solution or sonicating the tissue in either water or a fungicide solution before sterilization.

After washing, the explant tissue is immersed in an antiseptic solution to kill contaminants present on the surface. A 10–50% solution of commercial Clorox or laundry bleach (0.5–5.25% sodium hypochlorite) is one of the most commonly used sterilants in plant tissue culture. Other disinfectants that are used include ethanol, calcium hypochlorite, hydrogen peroxide, silver nitrate, bromide water, and mercuric chloride (see Table 1.2 for concentrations and sterilization times). The addition of a detergent such as polyoxyethylene sorbitan monolaurate (Tween-20) to the sterilant may enhance the effectiveness of the disinfectant by breaking the surface tension between the water and the plant tissue. The various sterilants may be used alone or in sequence to obtain the most effective sterilization procedure.

Following disinfestation or sterilization, the explant tissue must be rinsed several times in sterile distilled water to remove any remaining traces of the disinfectant. The damaged tissue ends of the explant are then removed and the explant subdivided into appropriate sizes. The explant is then placed on a nutrient medium designed for maximum growth of that particular species.

Plant tissues having internal contaminants can sometimes be disinfected by adding 10 mg/liter of benomyl or benlate to the culture medium or by treating the tissue with either benomyl or benlate before disinfestation with commercial bleach. Standard disinfestation techniques generally cannot remove internal contaminants. Antibiotics also may be added to the culture medium to check the growth of bacteria; however, antibiotics may be harmful to the particular species you are attempting to culture, so caution in their use must be exercised. One antibiotic, Cefotaxine (Sigma Chemical), has been used with excellent success in recent studies. Cefotaxine in potato culture media enhanced shoot proliferation in those cultures.

If contaminants are present, they usually become apparent within 3–5 days after inoculation. The appearance of contamination after approximately 10 days of culture generally indicates the presence of either an internal contaminant or the presence of mites or ants, which will spread contaminants rapidly through a culture room. The presence of contamination away from the explant tissue in the first several days of culture usually indicates sloppy culture techniques and not contaminated tissue.

Explants of some species frequently turn brown or black within several days after isolation; when this occurs, growth is generally inhibited and the tissue will die. Browning of tissue is most severe in species that contain high levels of tannins or other hydroxyphenols. Young tissues are less prone to browning than older tissues. The necrosis or browning results from the action of copper-containing oxidase enzymes (e.g., polyphenoloxidase and tryosinase), which are synthesized and/or released due to wounding during the excision and sterilization of the tissue (24). The browning of tissue can be prevented by (1) removing the phenolic compounds produced; (2) modifying the redox potential; (3) inactivating the phenolase enzymes; and (4) reducing the phenolase activity and/or substrate availability (6). In practice, the most feasible methods of inhibiting browning of the tissue are removing the phenolic substances produced and modifying the redox potential.

The phenolic compounds produced during the browning process can generally be removed in several ways. First, the tissue may be transferred frequently during the first 2–4 weeks in culture, so that large amounts of phenolic compounds do not build up. The interval between transfers may be 1–5 days, depending upon the quantity of phenolic compounds synthesized by a particular species. The transfer of tissues may be facilitated by culturing them on liquid medium instead of solid medium during the first several weeks. If this is done, the liquid medium containing phenolics can simply be siphoned off and fresh medium added with little disturbance to the tissue. Secondly, phenolics can be bound to compounds such as activated charcoal or polyvinylpyrrolidone (PVP); when bound, phenolics do not inhibit plant tissue. Charcoal is generally added to the culture medium at a concentration of 0.5–5.0 g/liter, while PVP is used at a concentration of 0.01–2%. When activated charcoal is used, care must be taken to ensure that the charcoal is suspended throughout the medium, as it tends to settle after autoclaving.

Tissue browning also can be prevented or reduced by lowering the redox potential with reducing agents or antioxidants. Tissues prone to browning are dipped into a sterile antioxidant solution immediately after excision. Compounds used as antioxidants include ascorbic acid, citric acid, L-cysteine, hydrochloride, 1,4-dithiothreitol, glutathione, and mercaptoethanol. Ascorbic acid and citric acid are the most commonly used antioxidants and generally are used together at concentrations of 50–150 mg/liter.

STAGE II—MULTIPLICATION OF THE TISSUE

The major goal of Stage II is the rapid multiplication of propagules. The plant material from Stage I is repeatedly subcultured in Stage II until the desired number of propagules or plantlets are obtained. Three procedures used for the "clonal" multiplication of plants during Stage II are described in this section.

Somatic Cell Embryogenesis

The process with the greatest potential for achieving rapid clonal micropropagation is somatic cell embryogenesis. In this process, a single cell is induced to produce an embryo, which in turn produces a complete plant. Somatic embryos are simply organized structures that originate from somatic cells but whose morphology resembles that of a developing zygotic embryo. Somatic embryogenesis has been successfully demonstrated in several plant species including carrot, *Antirrhinum*, *Petunia*, and many other species. Yields of 500 carrot embryos per gram of callus per month are not unusual (*17*). Somatic embryos may develop either in callus cultures or more commonly in cell suspension cultures.

Somatic embryogenesis in cell suspensions involves the following steps: (1) establishment of an actively growing stock callus on a medium containing a reduced nitrogen source (e.g., NH_4NO_3) and an auxin such as 2,4-D; (2) development of actively growing cell suspension cultures on a medium similar to that of the callus medium; (3) removal of the auxin or a reduction in auxin concentration; and (4) plating of the somatic embryos and their subsequent germination and plantlet development. The process of somatic embryogenesis is reviewed in depth by Kolenbach (*22*), Naraganaswamy (*30*), Raghaven (*31*), and Ammirato (*1*).

Enhanced Axillary Development

Axillary and terminal buds may be induced to develop *in vitro* by enhancing the development of quiescent or active shoot buds present. An explant containing a single bud may, depending upon the species and culture medium, develop into a single shoot or produce multiple shoots. As the new shoots develop, they in turn produce buds along their axis; through repeated subculture, this process can be repeated indefinitely. Callus formation may occur in association with bud development and adventitious shoots may originate from meristematic regions within the callus, thus producing more plantlets. Cytokinin concentrations of 1–30 mg/liter are used to enhance axillary bud proliferation. After several subcultures, the *in vitro*-derived plantlets can be transferred to Stage III to enhance the rooting of the material.

Although this is the slowest method of micropropagation, it is the most widely applicable in terms of the number of genera that can be propagated in this manner. Perhaps the greatest potential of this procedure is with woody plant species in which somatic embryogenesis and adventitious bud/shoot proliferation have not been very successful.

Adventitious Shoot Development

Adventitious shoots and related organs are structures that originate in tissues located in areas other than leaf axils or shoot tips. Adventitious shoots, roots, bulblets, and other specialized structures may originate from stems, leaves, tubers, corms, bulbs, or rhizomes. Adventitious shoots or organs may also originate from callus, which serves as an intermediate between the explant and plantlet production. The number of propagules is increased by subdividing and reculturing the in vitro-derived organs or callus. The plantlets or propagules attained using these techniques can then be transferred to Stage III for rooting.

The use of callus cultures and adventitious shoot proliferation may result in a loss of morphogenetic potential of the tissue or an increase in genetic variability. Several explanations have been proposed to account for the loss of morphogenetic potential (15): (1) loss of the meristematic centers in a callus due to the centers being stimulated to produce shoots through repeated subculturing; (2) variations or reductions in endogenous hormone levels that cannot be replaced by any known constituents; and (3) accumulation of chromosome abnormalities such as changes in the ploidy level or rearrangement of the chromosomes.

STAGE III—*IN VITRO* ROOTING AND CONDITIONING

In vitro-derived shoots may be induced to produce roots either in vitro during Stage III or in vivo during Stage IV. With certain species, the best results are obtained when shoots from Stage II are transferred to a rooting medium in Stage III. Stage III is generally one generation in length and lasts 2–4 weeks during which the plantlets are rooted and conditioned to a certain degree.

Without a doubt, one of the major obstacles to the micropropagation of plants in vitro is the rooting and acclimatization (conditioning, hardening off) of the plantlets. In the past, the majority of all shoots were rooted in vitro during Stage III. Although this procedure is still widely followed, it is now known that plantlets can be rooted outside of the culture vessel, and this alternative procedure is becoming more common.

From a commercial standpoint, the induction of roots in vitro is an expensive, labor-consuming process, which may account for 35–75% of the total cost of plants propagated in vitro (7). Furthermore, it has been reported that root systems induced in vitro are seldom functional when transferred to soil because the roots frequently lack root hairs and are generally so fragile that they are damaged during transfer (7). Despite these disadvantages, in vitro rooting techniques are still commonly used in research and many commercial laboratories. Indeed, with some species, in vitro rooting techniques may be the only practical method of rooting plantlets. McComb and Newton (26), however, have described a technique that involves inserting the base of the shoot into a flexible polyurethane foam submerged in a liquid rooting medium. The plantlets rooted in this manner can be transferred directly to the soil without removing their support.

Few plants produce roots under the conditions used to encourage multiple shoot formation in Stage II. In most instances, the presence of cytokinins in Stage II media inhibits root formation; thus, a separate root-inducing medium must be used during Stage III. Among the factors that influence the rooting of plantlets are growth regulators, macronutrients, micronutrients, organic supplements, support medium, light, and temperature.

In some species, all that is required to induce rooting is to transfer shoots or shoot clusters to a cytokinin-free medium. Species that may be taken directly from culture and rooted quite successfully without any further treatment include Aechmea (20), hybrid cherry (25), and azalea (9). However, in many species the initiation of roots occurs only in the presence of auxin. The auxins (see Fig. 2.1) most frequently incorporated into media to induce rooting are IAA (0.1–10.0 mg/liter), NAA (0.05–1.0 mg/liter), and IBA (0.5–3.0 mg/liter). The response of shoots to these auxins and the formation of roots in their presence is very species dependent. Occasionally sufficient cytokinin may be carried over from Stage II to inhibit rooting during Stage III.

Roots initiated in the presence of auxin may fail to continue to grow in its presence and will only resume growth once the auxin source is removed. The need for preculturing the shoots on an auxin source for root initiation may be eliminated by dipping the shoots for a short period of time in a sterile, concentrated auxin solution before reculturing on the Stage III medium.

Concentrations of macro- and micronutrients are frequently reduced to half their normal values during the rooting phase, although this varies with species. The favorable effect of reduced concentrations of macro- and micronutrients is probably due to a decreased requirement of the plantlets for nitrogen. Although organic supplements may be included in the rooting medium, they do not appear to be essential. Kamada and Harada (21) noted that amino acids and vitamins could either stimulate or inhibit rooting, depending upon species. The presence of sugar has been found essential for the in vitro rooting of many species. Gautheret (10) demonstrated that rooting was better when glucose, rather than fructose or sucrose, was the carbohydrate source.

Hyndman et al. (16–18) reported that in roses the number of roots formed per shoot progressively increased as the sucrose concentration increased from 1 to 7%, but 9% sucrose was supraoptimal.

The conditions of the cultural environment may also influence the in vitro rooting of shoots. It has been noted that root hairs do not normally grow on roots that develop in agar, presumably due to poor aeration. Root hairs are produced on roots when the shoots are rooted in a liquid medium, and more root hairs are formed when the shoot is supported by a paper bridge above the liquid medium. Light may be required for the initiation of roots in certain species. This lighting requirement may reflect the need of the plant for a specific level of endogenous carbohydrates for rooting to occur. The carbohydrates may be supplied by the culture medium or by photosynthesis, but if the carbon reserves fall below the critical level, rooting will not occur. The use of light during the rooting phase may not be beneficial in other species and may even be inhibitory in some. Exposure of plantlets to a short period of darkness in the presence of auxin followed by rooting under a light regime may be required for other species. Lane (23) and Hammerschlag (13, 14) have both shown that root formation is generally enhanced at higher temperatures, even for plants from normally cool environments. The temperature that appears to stimulate rooting best generally is 25°–28°C.

Several researchers have reported that rooting of in vitro-derived plantlets can be enhanced by combining techniques of both in vitro and in vivo rooting. Murashige et al. (29) reported poor rooting in gerbera when microcuttings were dipped in NAA, IAA, or IBA and then directly inserted into the soil mix, but enhanced rooting was observed when the cuttings were cultured on a medium containing 10 mg/liter IAA for 10 days before transfer to a soil mix. Debergh and Maene (7) improved rooting of Begonia tuberhybrida shoots by culturing the shoots for 10 days in petri dishes containing a 2-mm layer of 2 mg/liter IBA in distilled water before inserting the shoots into a peat compost. Zimmerman and Fordham (39) obtained enhanced root proliferation by culturing in vitro-derived shoots in a liquid medium in the dark for 3–7 days before inserting them into preformed peat plugs. The medium used in their study consisted of 43.8 mM sucrose and 1.5 μM IBA. The addition of half-strength MS salts or of phloroglucinol to the rooting medium did not enhance rooting; IBA was the most effective auxin source evaluated, followed by NAA and IAA. These authors reported that small visible root initials were present on the base of the cuttings after 7 days of dark incubation and that these roots became well developed, often with secondary branching, after 3 weeks in the peat plugs (39).

In general, shoots that are normally subcultured as clumps during the multiplication phase must be separated into individual shoots before being placed on the rooting medium. However, better rooting of strawberry transplants was achieved if the shoots were transferred to the rooting mix as an unseparated clump rather than as individual plantlets. On the other hand, Murashige et al. (29) noted that single gerbera shoots were more prone to initiate roots than were unseparated shoot clusters.

STAGE IV—*IN VIVO* ROOTING AND ACCLIMATIZATION

When shoots are transferred outside the culture vessel for rooting, they must be placed in some form of medium to support their growth. To avoid confusion with the media used for tissue culture, such artificial soil mixtures will be referred to as *rooting mixtures* or *soil mixtures*. Rooting mixtures generally include materials such as peat, bark, perlite, vermiculite, pumice, sand, and soil and may be supplemented with a small amount of lime or fertilizer. Marked differences in root formation with these various materials may be observed. Peat may be too acid for the rooting of certain species, whereas vermiculite may be too alkaline. The ideal rooting mixture has a neutral or slightly acidic pH and high water-holding capacity, yet provides good drainage and aeration.

Acclimatization has been defined as the process by which an organism adapts to environmental change (4). Acclimatization is necessary because in vitro-derived plantlets are not adapted nor suited for in vivo conditions. Plantlets that are to be acclimatized should be well-proportioned in regards to roots and shoots. If the plantlets have been rooted in vitro, they generally must be transplanted to a rooting mixture and maintained under partial shade and high relative humidity for several days. A suitable environment generally can be created by placing the plantlets in a clear plastic bag or box, or under intermittent water mist. Plants are acclimatized by gradually reducing the relative humidity in their environment. This may be accomplished by simply reducing the amount of mist plantlets receive or by gradually slitting or opening the plastic bags or boxes that hold the plantlets. Several workers have demonstrated that spraying plantlets with antitranspirants is beneficial in hardening off plants. Wardle et al. (37, 38) reported that cauliflower plantlets sprayed with antitranspirants had a two to three times greater cuticular resistance to water loss and required less attention during the acclimatization process than did untreated controls. Similar results were reported with kangaroo paws by McComb and Newton (26).

Plantlets rooted in vitro may actually harden off somewhat in the culture vessel before transfer to the rooting mix. This may occur if the closures are removed from the culture vessels and plantlets are left on the nutrient medium for an additional week or two.

The treatment of newly transplanted plantlets with fungicides is another practice that may increase the survival of plants. Several commercial rooting powders contain a fungicide (e.g., Thiram), which may account for the increased survivability of plantlets treated with such powders.

Several factors—including photosynthesis rate and cuticular wax on the leaves—have been associated with low survivability of in vitro-derived plantlets. The internal structure and anatomy of in vitro-propagated plants is often initially different from that of greenhouse or field-grown plants. This change in anatomy or ultrastructure may affect the photosynthetic process within the plants. Plantlets grown in vitro on a carbohydrate-enriched medium generally produce only a very small part of their carbohydrate requirement through photosynthetic CO_2 fixation. When the plantlets are transferred to in vivo conditions, they must become fully autotropic. Grout and Aston (11) noted that the rate of CO_2 fixation in cauliflower plantlets 7 days after transplantation was higher than CO_2 fixation in cauliflower plantlets in vitro, but the transplants were still releasing more CO_2 than they were assimilating through photosynthesis. Dark respiration of the plantlets was also higher than that of seedlings of a similar age, suggesting that the newly transferred plantlets were using a large amount of energy in adapting to their new environment. A positive carbon balance in the cauliflower plantlets was observed 14 days after transplanting. The survival of newly transferred plantlets largely depends upon large starch reserves that accumulate during culture.

In vitro-derived plantlets have much less cuticular wax on the surface of their leaves than do similar plants grown in a greenhouse or the field. This reduction in cuticular wax causes plantlets to lose water more rapidly than normal plants and may contribute to the poor survivability of plantlets (34, 35). The high humidity present in the culture vessel may be responsible for the reduction in wax on the leaf surfaces. However, Sutter and Langhans (34, 35) demonstrated that both high humidity and low light levels tended to inhibit wax formation. Cuticular wax is not reduced on in vitro shoots in all species, and the survivability of plantlets in the greenhouse does not always correlate with the presence or absence of cuticular wax (34, 35). Another cause for water loss from the leaf surface is the ineffective control of the stomata. Wardle et al. (37, 38) reported that stomata on cauliflower leaves from in vitro-derived plantlets were unable to close in response to ABA. It has been mentioned by some that the leaves on *in vitro*-derived plantlets may serve as little more than storage organs, since they grow only slightly after transplanting and die back quickly (*37, 38*).

ABNORMALITIES IN *IN VITRO*-DERIVED PLANTLETS

A wide degree of variation in plantlets produced *in vitro* have been reported. These changes are the result of the presence of chimeras, changes in chromosome number, rearrangement in chromosomes, reversion of the material back to the juvenile stage, and nuclear and cytoplasmic mutations. Some phenotypic changes observed in *in vitro*-derived plants include the presence or absence of pubescence, changes in leaf morphology, dwarfing, loss of pigmentation, and alteration in flower morphology. While these changes are undesirable for the propagator trying to clonally propagate these plants, they may provide the plant breeder with a wider genetic base containing certain desirable traits. Many of the observed abnormalities appear to have an epigenetic or physiological basis and therefore are reversible.

Abnormal leaf shapes or arrangements are one of the easiest changes to recognize. Leaves may lack trichomes, or the leaf shape may be altered. Leaf width is initially smaller in some grass lines grown *in vitro*; changes in leaf arrangement (e.g., alternate vs. opposite) have been reported in other species. Variegation in leaves may appear in certain species and disappear in others when plants are propagated *in vitro*.

Anderson *et al.* (*2*) have demonstrated that some varieties of strawberries when propagated *in vitro* yield plantlets with multiple apices. The presence of faciated shoot apices has also been reported in other species such as carnation (*12*) and *Celosia* (*8*).

Reduction or enhancement in plant vigor may also occur in plants propagated through tissue culture. Bush *et al.* (*5*) reported that chrysanthemum plants regenerated from petals lacked the vigor of plantlets derived from shoot apices. Bilkey and Cocking (*3*) reported that plantlets derived from the internal pith tissue of leaf petioles of *Saintpaulia* had more vigor than those derived from the epidermis. Croughan (*6*), on the other hand, reported that rice plants produced *in vitro* were dwarfed in nature but produced larger yields compared with seed-produced controls.

Albinism is another abnormality that often occurs during *in vitro* propagation. The production of albino plants during tissue culture is fairly common in the cereals and grasses but has also been reported in *Citrus* (*32*) and *Pyrus* (*26*). It has been suggested that the production of albino plantlets might be due to the disorganization of the plastids *in vitro*, while others have noted deficiencies in certain ribosomal RNAs in albino shoots. The addition of cytokinins to the culture medium may decrease the chance of albino plant production in grasses.

The causes of many of the abnormalities observed in *in vitro*-derived plants can be traced to constituents used in the culture medium. Cytokinins required for shoot induction and multiplication may cause abnormalities such as decreased rooting, stunted or compact plants (19, 33), increased branching (19, 33), or slender stems and leaves. Abnormalities in regenerated plants have been less frequently attributed to auxins; however, abnormal fertility of the flowers, abnormal petal shapes, decreased vigor, malformed leaves, and increased lateral shoot formation have all been attributed to improper auxin concentrations. Other factors that may result in the production of abnormal plantlets if not closely monitored include GA_3, temperature, osmotic potential of the medium, and agar concentration.

REFERENCES

1. Ammirato, P. V. 1983. Embryogenesis. pp. 82–123. *In* Handbook of Plant Cell Culture, Vol. 1. D. A. Evans, W. R. Sharp, P. V. Ammirato, and Y. Yamada (editors). MacMillan, New York.
2. Anderson, H. M., A. J. Abbott, and S. Wiltshire. 1982. Micropropagation of strawberry plants *in vitro*—effect of growth regulators on incidence of multi-apex abnormality. Sci. Horticulturae 16:331–341.
3. Bilkey, P. C. and E. C. Cocking. 1981. Increased plant vigor by *in vitro* propagation of *Saintpaulia ionantha* Wendl. from sub-epidermal tissue. HortScience 16(5):643–644.
4. Brainerd, K. E. and L. H. Fuchigami. 1981. Acclimatization of aseptically cultured apple plants to low relative humidity. J. Am. Soc. Hort. Sci. 106(4):515–518.
5. Bush, S. R., E. D. Earle, and R. W. Langhans. 1976. Plantlets form petal segments, petal epidermis and shoot tips of the periclinal chimera, *Chrysanthemum mortifolium* 'Indianapolis'. Am. J. Bot. 63(6):729–737.
6. Croughan, T. P. 1984. Personal communication.
7. Debergh, P. C. and L. J. Maene. 1981. A scheme for commercial propagation of ornamental plants by tissue culture. Sci. Horticulturae 14:335–345.
8. Driss-Ecole, D. 1981. Fasciation d'extremites caulinaries du *Celosia cristata* (Amarantacees) cultivees *in vitro*. Can. J. Bot. 59:1367–1372.
9. Economou, A. S., P. E. Read, and H. M. Pellet. 1981. Micropropagation of hardy decidious azaleas. HortScience 16(3):452.
10. Gautheret, R. J. 1969. Investigation on the root formation in the tissues of *Helianthus tuberous* cultured *in vitro*. Am. J. Bot. 56(7):702–717.
11. Grout, B. W. W. and M. J. Aston. 1978. Transplanting of cauliflower plants regenerated form meristem culture. II. Carbon dioxide fixation and the development of photosynthetic ability. Hort. Res. 17:65–71.
12. Hackett, W. P. and J. M. Anderson. 1967. Aseptic multiplication and maintenance

of differentiated carnation shoot tissue derived from shoot apices. Proc. Am. Soc. Hort. Sci. 90:365–369.

13. Hammerschlag, F. 1982. Factors influencing *in vitro* multiplication and rooting of the plum rootstock Myrobalan (*Prunus cerasifera* Ehrh.). J. Am. Soc. Hort. Sci. 107:44–47.

14. Hammerschlag, F. 1981. *In vitro* propagation of Myroblam plum (*Prunus cerasifera*). HortScience 16(3):283.

15. Hughes, K. W. 1981. Ornamental species. pp. 5–50. *In* Cloning Agricultural Plants via *In Vitro* Techniques. B. V. Conger (editor). CRC Press, Boca Raton, Fla.

16. Hyndman, S. E., P. M. Hasegawa, and R. A. Bressan. 1981. Sucrose and nitrogen regulation of adventitious root initiation from cultured rose shoot tips. HortScience 16(3):463.

17. Hyndman, S. E., P. M. Hasegawa, and R. A. Bressan. 1982. Stimulation of root initiation from cultured rose shoots through the use of reduced concentrations of mineral salts. HortScience 17(1):82–83.

18. Hyndman, S. E., P. M. Hasegawa, and R. A. Bressan. 1982. The role of sucrose and nitrogen in adventitious root formation on cultured rose shoots. Plt. Cell, Tissue and Organ Culture 1:229–238.

19. Jones, J. B. 1979. Commercial use of tissue culture for the production of disease-free plants. pp. 441–452. *In* Plant Cell And Tissue Culture. W. R. Sharp, P. O. Larsen, E. F. Paddock, and V. Raghaven (editors). Ohio State Univ. Press, Columbus.

20. Jones, J. B. and T. Murashige. 1974. Tissue culture propagation of *Aechmea fasciata* baker and other bromeliads. Comb. Proc. Plt. Prop. 24:117–126.

21. Kamada, H. and H. Harada. 1979. Influence of several growth regulators and amino acids on *in vitro* organogenesis of *Torenia fournieri* Lind. J. Exp. Bot. 30(114):27–36.

22. Kolenbach, H. W. 1977. Basic aspects of differentiation and plant regeneration from cell and tissue cultures. p. 355. *In* Plant Tissue Culture and its Bio-Technological Applications. W. Barz, E. Reinhard, and M. H. Zenk (editors). Springer-Verlag, New York.

23. Lane, W. D. 1978. Regeneration of apple plants from shoot tips. Plt. Sci. Lett. 13:281–285.

24. Lerch, K. 1981. pp. 143–186. *In* Metal Ions in Biological Systems. H. Sigel (editor). Marcel Dekker, New York.

25. Lineberger, R. D. 1982. *In vitro* propagation of 'Holly Jolivette' cherry. HortScience 17(3):533.

26. McComb, J. A. and S. Newton. 1981. Propagation of kangaroo paws using tissue culture. 56(2):181–183.

27. Mehra, P. N. and K. Jaidka. 1979. In vitro morphogenetic studies in pear (*Pyrus communis*). Phytomorph. 29:286–298.

28. Murashige, T. 1974. Plant propagation through tissue culture. Ann. Rev. of Plt. Physiol. 25:135.

29. Murashige, T., M. Serpa, and J. B. Jones. 1974. Clonal multiplication of Gerbera through tissue culture. HortScience 9(3):175–180.

30. Naraganaswamy, S. 1977. Regeneration of plant tissue culture. pp. 179–202. *In* Applied and Fundamental Aspects of Plant Cell, Tissue and Organ Culture. J. Reinert and Y. P. S. Bajaj (editors). Springer-Verlag, New York.

31. Raghaven, V. (editor). 1976. Experimental Embryogenesis in Vascular Plants. Academic Press, New York.

32. Raj Bhansali, R. and H. C. Arya. 1979. Organogenesis in *Citrus limethioides* (Sweet Lime) callus culture. Phytomorph. 29:97–100.

33. Smith, R. H. and A. E. Nightingale. 1979. *In vitro* propagation of *Kalanchoe*. Hort-Science 14(1):20.

34. Sutter, E. and R. W. Langhans. 1978. Epicuticular wax and cuticle formation in meristem-regenerated plantlets of carnation. HortScience 13(3):348.

35. Sutter, E. and R. W. Langhans. 1979. Epicuticular wax formation on carnation plantlets regenerated from shoot tip cultures. J. Am. Soc. Hort. Sci. 104(4):493–496.

36. Takatori, F. H., T. Murashige, and J. I. Stillman. 1968. Vegetative propagation of asparagus through tissue culture. HortScience 3(1):20–22.

37. Wardle, K., A. Quinilan, and I. Simpkins. 1979. Absicisic acid and the regulation of water loss in plantlets of *Brassica oleracea* L. var. *botrytis* regenerated through apical meristem culture. Ann. Bot. 43:745–752.

38. Wardle, K., V. Dalsou, I. Simpkins, and K. C. Short. 1983. Redistribution of rubidium in plants of *Chrysanthemum morifolium* Ram. cv. Snowdown derived from tissue cultures and transferred to soil. Ann. Bot. 51:261–264.

39. Zimmerman, R. H. and I. Fordham. 1985. Simplified method for rooting apple cultivars *in vitro*. J. Amer. Soc. Hort. Sci. 110:34–38.

4

Application of Tissue Culture Techniques to Horticultural Crops

ORNAMENTALS

The first and, to date, the most extensive practical application of tissue culture techniques to horticultural crops involves the multiplication of ornamental plant species. The early work by Morel (4) with orchids, the development of the Murashige and Skoog medium (5), and the efforts of Murashige (6, 7) to enhance the practical side of tissue culture science stimulated the development and widespread use of tissue culture as a means of micropropagating ornamentals. Herbaceous ornamentals have adapted to tissue culture techniques with relative ease; more attention and efforts have been required to successfully apply *in vitro* culture techniques to woody plant species.

There have been three primary goals in using *in vitro* culture techniques with ornamentals: (1) the elimination of diseases and the production of disease-free plant material; (2) the rapid production of a large number of genetically identical plants, and (3) introduction of new varieties and/or genotypes. The majority of the tissue culture research with ornamentals has centered around the first two goals. Several commercial facilities in the United States are producing *Chrysanthemum* and *Dianthus in vitro*, and numerous commercial operations are involved in the mass propagation of ornamentals through tissue culture techniques.

Tissue culture techniques offer several advantages for a commercial operation in addition to the goals listed above. Stock plants for micropropagation may not be required once a propagation scheme is established, since the explants for future propagation can be derived from plantlets growing *in vitro*. Tissue culture techniques can also be used to obtain hybrids from incompatible species through either embryo rescue techniques, ovule culture, embryo culture, or somatic hybridization. The production of haploid plants through tissue culture may also help plant breeders obtain homozygous, pure-breeding plants.

Reviews on the micropropagation of ornamentals have been written by Murashige (6, 7), Holdgate (2), Hughes (3), and George and Sherrington (1).

TISSUE CULTURE OF FRUIT CROPS

Most fruit-bearing species have traditionally been propagated using vegetative techniques so that desirable traits are preserved. Micropropagation techniques may be used to enhance or, in the future, may replace the vegetative propagation techniques now being employed. In certain regions of the world, this is already occurring. For example, the majority of all commercially produced strawberries are produced from *in vitro*-derived plantlets in certain European countries.

The potential impact of tissue culture on the fruit industry has been discussed thoroughly in the literature in recent years. Most tissue culture programs have centered around one of the following areas: (1) mass propagation of desirable lines and rootstocks, (2) elimination of viruses from plant tissues, (3) rapid micropropagation of desirable lines from breeding programs, (4) preservation of germplasm, and (5) production of haploids for plant breeding.

The fruit crops that have received the most attention in regards to tissue culture include strawberries, apples, peaches, grapes, blueberries, thornless blackberries, and red raspberries. A survey of the literature indicates that most investigators are using a modification of the Murashige and Skoog medium. Most research has focused on (1) the induction and maintenance of callus; (2) the production or enhancement of adventitious and axillary shoots, buds, or organs; or (3) the rooting of *in vitro*-derived plantlets. The production of callus for most fruit species has been detrimental; however, adventitious plantlet formation has been stimulated from callus of certain species such as strawberry.

TISSUE CULTURE OF VEGETABLE CROPS

Vegetables as a group constitute one of the largest agricultural commodities in the world. Vegetables are among the most beneficial plants in terms of human nutrition, as they supply necessary vitamins, minerals, and proteins to human diets. The primary goals of the *in vitro* propagation of vegetable crops include (1) production of large number of plantlets from species in which plant development from seed is difficult, (2) clonal propagation of large numbers of genetically identical plantlets, (3) production of virus-free plant material, and (4) crop improvement through various techniques of genetic modification.

All three in vitro multiplication methods discussed in Chapter 3 have been used successfully with vegetable crops. For example, in cauliflower and potato, somatic embryos may arise directly from the explant tissue. Somatic embryogenesis may also occur from callus cultures of asparagus, carrots, celery, potato, sweet potatoes, and pumpkins or from cell suspension cultures of carrots.

Another method of propagating vegetable crops is through enhanced axillary branching using stem tips and lateral buds as the explants. The advantage of this type of micropropagation is that very little callus is formed and the degree of genetic abnormalities is often reduced. Once the explants are established and axillary bud development enhanced, the cultures can be subcultured for many generations, resulting in increased shoot formation. Shoots can be excised after elongation and generally rooted either in vitro or in a growth chamber or greenhouse environment. Vegetable crops that have been micropropagated using these techniques include asparagus, broccoli, brussels sprouts, cauliflower, cucumber, cabbage, garlic, lettuce, tomato, potato, and sweet potato.

Adventitious shoot formation also has been used to propagate vegetable crops in vitro. Lettuce and cabbage are examples of vegetable crops in which adventitious plantlets have originated directly from the primary explant. Adventitious plantlet formation from callus has been reported with asparagus, broccoli, brussels sprouts, chive, cabbage, carrot, garlic, kale, lettuce, pepper, potato, tomato, and sweet potato. The disadvantage of adventitious plantlet formation is that genetic variability often increases, especially when the plantlets are derived from callus. This variability generally tends to increase as the duration of a generation or the length of time callus is in culture increases. The genetic variability commonly observed in these cultures includes variation in phenotypic expression, yield variability, and loss of organogenic potential and is generally the result of chromosome abnormalities and/or ploidy changes in chromosome number.

REFERENCES

1. George, E. F. and P. D. Sherrington. 1984. Plant Propagation by Tissue Culture. p. 703. Exgetics Ltd., Eversley, England.
2. Holdgate, D. P. 1977. Propagation of ornamentals by tissue culture. pp. 18–43. In Applied and Fundamental Aspects of Plant, Cell, Tissue and Organ Culture. J. Reinert and Y. P. S. Bajaj (editors). Springer-Verlag, New York.

3. Hughes, K. 1981. Ornamental species. pp. 5–50. *In* Cloning Agricultural Plants via *In Vitro* Techniques. B. V. Conger (editor). CRC Press, Boca Raton, Fla.
4. Morel, G. M. 1960. Producing virus-free *Cymbidium*. Am. Orch. Soc. Bull. 29:495.
5. Murashige, T. and F. Skoog. 1962. A revised medium for rapid growth and bioassays with tobacco cultures. Physiol. Plant. 15:473.
6. Murashige, T. 1974. Plant propagation through tissue culture. Annu. Rev. Plant Physiol. 25:135.
7. Murashige, T. 1977. Clonal crops through tissue culture. pp. 392–403. *In* Plant Tissue Culture and its Bio-Technological Application. W. Barz, E. Reinhard, and M. H. Zenk (editors). Springer-Verlag, New York.

PART II

Callus (Tissue) and Organ Culture

5

Overview of Callus (Tissue) and Organ Culture

The term *callus* refers to tissue arising from the disorganized proliferation of cells from segments (explants) of plant organs. Callus formed during *in vitro* culture has some similarities to tissue arising *in vivo* after injury to plants (so-called wound callus). However, there often are differences in morphology, cellular structure, growth, and metabolism between callus derived through tissue culture and natural wound callus.

The technique of callus (tissue) culture was first developed in the late 1920s and early 1930s and was one of the primary methods of tissue culture for many years. Very few plant organs fail to respond to treatments designed to induce callus formation. Partly for this reason, callus cultures are very frequently used in a variety of biochemical, physiological, and genetic experiments.

SOURCE OF MATERIAL

The isolation and successful establishment of callus cultures depends primarily on the culture conditions employed, not on the source of plant material. Callus cultures have been successfully initiated from both dicotyledonous and monocotyledonous plants. Among the tissues that have been induced to produce callus under *in vitro* culture conditions are vascular cambia, storage organs, pericycle of roots, endosperm, cotyledons, leaf mesophyll, and provascular tissue.

STERILIZATION OF SOURCE MATERIAL

The outer surfaces of plants grown under natural, greenhouse, or growth chamber conditions generally are infested with spores and other microbial cells. Internal tissue is generally pathogen free; however, mature internal vascular tissue may be infected with vascular fungi or bacteria without showing any visible symptoms. It is, therefore, very important that only healthy tissue be used for tissue culture purposes. Before any explant can be inoculated, it is essential to destroy all microorganisms. Even if the surface layers of the explant are to be removed (e.g., by dissecting out and using an apical dome as the explant), it is necessary to first sterilize the surface layers to prevent carrying pathogens into the inner tissues or onto the instruments.

Procedures for sterilizing plant materials were discussed in Chapter 1 ("Sterilizing Plant Materials") and in Chapter 3 ("Stage I—Establishment of an Aseptic Culture"). Techniques described at the end of this chapter will give you practice in the common techniques for surface-sterilizing plant parts.

PREPARATION OF THE EXPLANT

The size and shape of the initial explant is normally not critical, although proliferation may fail to occur if the explant is below a critical size. In general, fairly large pieces of material are used because the large number of cells present increases the chances of obtaining a viable culture. Leaves, stems, flowers, anthers, fruit, seed, apical regions, and roots have all been used to establish callus cultures.

CULTURE MEDIUM

A continuous supply of inorganic salts, vitamins, sucrose, and growth regulators are required to sustain an actively growing callus culture. All callus cultures require the nutrients discussed in Chapter 2.

The response of an excised piece of tissue, providing all other elements are at optimum levels, depends largely on the concentration of auxins and cytokinins. Tissue response depends on both the endogenous level of these growth regulators within the tissue and on the exogenous supply in the medium. The auxins normally used for initiation and maintenance of callus cultures include IAA at 10^{-5} to 10^{-10} M and NAA at 10^{-5} to 10^{-10} M. Some tissues may fail to initiate callus in response to auxin alone and may require the presence of a cytokinin for callus growth. Kinetin at 10^{-7} to 10^{-6} M has proven very effective in enhancing IAA callus initiation from tobacco pith cultures.

The maintenance of callus may require the presence of various amino acids in the medium. Those commonly used to support callus growth include glycine, arginine, or mixtures of amino acids such as those found in casein hydrolysate. The formulations of many common media are given in Chapter 2.

Callus cultures generally are grown on a solid medium, although a liquid medium sometimes is used. Many of the early studies with callus cultures were carried out on media solidified with agar, gelatin, or silica gel. More recently, certain types of acrylamide gels have been used for solidifying media. Agar is the most popular gelling agent and is generally included at 6–10 g/liter. One advantage of using an agar solidified medium is that when combined with water it forms a gel that melts at approximately 100°C and then solidifies at approximately 45°C. This results in a gel that is solidified and stable at most incubation temperatures. Additionally, agar gels do not react with the media constituents and are not digested by plant enzymes. Refer to "Solidifying Agents or Support Systems" in Chapter 2 for further discussion.

Solid cultures have several disadvantages that make liquid cultures preferable in some cases. For example, only part of a callus culture is actually in contact with the surface of the medium; therefore, inequalities in growth may arise due to gradients in nutrients, in the exchange of gases, and in toxic waste products that may develop between the callus and the medium. In addition, the callus may be subject to polarization by gravity and variations in light intensity, which may affect growth. The growth of callus in solid culture also may be limited in certain directions by the medium or walls of the glassware. Finally, callus grown in a solid medium cannot be transferred to liquid medium without some disturbance to the tissue occurring. Despite these limitations, solid culture still remains the method of choice for routine maintenance of callus cultures.

Use of stationary cultures of tissue in an *unshaken* liquid medium has the advantages of solid culture as well as the advantage that many inhibitory compounds commonly contained in solidifying agents are not present. This method of growing cultures is generally reserved for work on mineral nutrition. The tissue is generally placed on filter disks held at the interface of the medium with the air. The filter paper acts as a wick providing nutrients while keeping the tissue on the gaseous phase.

SUBCULTURING

Callus cultures must be transferred periodically to fresh nutrient medium. Extensive growth of the cultures leads to a buildup of waste products, exhaustion of the nutrients, and drying out of solid media.

Cultures incubated at 25°C or above should be subcultured every 4–6 weeks. Most callus cultures also require frequent subdivision and transfer of the separated pieces to fresh medium unless the callus is in an early stage of development. Small healthy looking pieces should be transferred, but the pieces should not be too small, as they may not survive the transfer. Failure to subdivide and transfer the callus leads to necrosis followed by death of the callus. Callus pieces derived from necrotic tissue tend to grow much less actively than pieces taken from actively growing tissue. Active growth from necrotic tissue can sometimes be restored by frequent subculturing. Loss of vigor may be attributed to a carryover of toxic materials in the callus from one generation to the next. Viral contamination also may result in decreased vigor.

Although embryonic callus may appear to be slower growing than nonembryonic callus, it is generally as healthy, somewhat less friable, and much more responsive to regeneration than nonembryonic tissue.

MEASUREMENT OF CALLUS GROWTH

Analysis of callus cultures is usually based on fresh weight or dry weight measurements. Measurement of fresh weight is a quick and easy method for assessing callus growth and does not involve destruction of the material. However, fresh weight measurements mainly reflect the water content of the tissue and may not accurately reflect the actual growth of the callus. Other measures that are used to assess callus growth include callus dry weight, cell number of the aggregate, mitotic index of the callus, and respiration rate. The procedure for determining the cell number in explants and callus is given in Appendix 10; instructions for dry weight determinations are presented in Appendix 11.

Techniques for Establishing Plant Tissue in Culture
As has been emphasized several times already, obtaining and maintaining sterile conditions is essential in tissue culture work. Sterilization (disinfestation) of an explant is generally accomplished by submerging the plant in a strong disinfectant solution for a short period of time, then rinsing away the toxic residue with sterile water. Some disinfectants used for this purpose, with suggested concentrations and exposure times, are listed in Table 1.2. The outer surfaces of many plant surfaces may contain a waxy cuticle layer. It may be necessary to add 2–3 drops/liter (0.01%) of a commercial detergent to the disinfectant to allow better penetration of the waxy layer and of the small pores on the tissue surface. Many dishwashing liquids or commercially prepared wetting agents such as polyoxyethylenesorbitan (Tween-20) are satisfactory for this purpose.

In this exercise, you will treat explants from three different source materials with several different sterilants, then inoculate the explants onto a solid medium and observe if and when contamination first appears. The effect of acidifying certain sterilants and of sterilization time also will be tested.

Materials Required
1. sterile plastic petri dishes (100 × 15 mm)
2. 11 1000-ml beakers
3. 2 scalpels with new blades, 3 spatulas, and 3 pairs of forceps
4. Waterproof pen and stick-on labels for labeling tubes and petri dishes
5. Bunsen or ethanol burner
6. Slant racks for holding tubes
7. 1 liter of saturated $CaCl_2$ solution
8. 1 liter of 5% Clorox solution
9. 1 liter of 20% Clorox solution
10. 1 liter of 50% Clorox solution
11. 250 ml of 1.0 N HCl
12. 1 liter of 95% ethyl alcohol
13. 1 liter of 50% ethyl alcohol
14. Leaf pieces of African violet
15. Seed of creeping bentgrass
16. Bulb scales of *Lilium longiflorum*

Procedures
1. Prepare 2 liters of water–agar medium by dissolving 6 g/liter of agar in 2000 ml of water. Adjust the pH of the medium to 5.7 ± 0.1 with NaOH or HCl. Bring the medium to a boil, autoclave, cool, then pour 15 ml of medium into each petri plate under aseptic conditions. Allow 1 hr for the medium to solidify.
2. To 500 ml of each of the following sterilants, add enough 1.0 N HCl to bring the pH to 6.5: saturated calcium chloride, 5% Clorox, 20% Clorox, and 50 Clorox. Prepare solutions under a fume hood and avoid breathing fumes.
3. Set up three sets of treatment tubes by dispensing 500 ml of the various sterilants into beakers according to the protocol given in Table 5.1 (label beakers).
4. Collect 22 healthy leaves from a *Saintpaulia ionantha* plant by removing the leaf along with approximately 5 cm of the petiole. Two leaves each will be used for the 11 treatments in this experiment. Place the leaves in the 11 1000-ml beakers and apply the sterilization treatments for the required 7 min.

Table 5.1.   Protocol for Establishing Three Sets of Treatment Tubes

	Duration (min)		
Treatment	*Lilium*	*Saintpaulia*	*Agrostis*
Water control	10	7	10
Saturated calcium chloride	10	7	10
Acidified saturated calcium chloride	10	7	10
5% Clorox	10	7	10
Acidified 5% Clorox	10	7	10
20% Clorox	10	7	10
Acidified 20% Clorox	10	7	10
50% Clorox	10	7	10
Acidified 50% Clorox	10	7	10
50% ethyl alcohol	10	7	10
95% ethyl alcohol	10	7	10

Collect 2 large *Lilium* bulbs from a local garden center or nursery. Remove the scales from the bulb under running water to remove any excess soil present on the scales. Place the bulb scales in a 1000-ml beaker and wash the scales under running water for another 5 min. Next, divide the scales into 11 groups making sure that bulb scales of all sizes are present in each of the 11 groups. Place each of these groups of bulb scales into the 11 1000-ml beakers and apply the sterilization treatments for the required 10 min.

Obtain approximately 100 g (about ¼ lb) of *Agrostis* seed or any other turfgrass seed from a local garden center or nursery. Divide the seed into 11 5-g samples and place one of each group into the 11 1000-ml beakers. Apply the sterilization treatments for the required 10 min.

5.   Administer the treatments as described in the preceding for each of the species. Acidified treatments should be handled in a fume hood. Once the tissue piece has been sterilized, rinse it three times with sterile distilled water. *Saintpaulia* leaves should be prepared by sectioning the leaves into 1-cm² sections with a sharp scalpel. *Lilium* bulb scales must be sectioned horizontally into 2-mm thick sections. *Agrostis* seed can be placed directly onto the culture medium. Enough plates should have been prepared so that 4 plates per treatment per plant species can be inoculated (i.e., there should be 44 plates, 4 per treatment, for *Saintpaulia*, *Agrostis*, and so on). Once the plates are inoculated, seal them with Parafilm, label, and incubate under low light at 25°C.

Scheduling

Event	Timing
Inoculation of explant	Day 0
First appearance of contamination	ca. Day 3–7
Termination of experiment	Day 14

Recording Results

1. Record all details of setting up the experiment.
2. Make visual observations every other day and record when contamination first appears.

REFERENCES

1. Basile, D. V. 1972. A method for surface sterilizing small plant parts. Bull. Torrey Bot. Club 99(6):313–316.
2. Street, H. E. 1973. Plant tissue and cell culture. Blackwell Publications, London.
3. Wetherell, D. F. 1982. Introduction to *in vitro* propagation. Avery Plant Tissue Culture Series. Avery Publishing Group, Inc., Wayne, NJ.

In Vitro Propagation of African Violets

The African violet, *Saintpaulia ionantha*, is propagated vegetatively from leaf cuttings and is grown on a large scale commercially and on a small scale by many home gardeners (4). When one or more shoots are allowed to develop on a cutting during vegetative propagation, constrictions are imposed by the multiplicity of plantlets in a limited growing space, resulting in unsymmetrical plants with elongated sideways displaced petioles. Propagation by tissue culture overcomes this problem and results in a large number of well-formed single-stemmed plants from a given amount of leaf tissue.

Bilkey et al. (1) demonstrated the high regenerative capacity of African violet petiole tissue when optimal levels of growth hormones, particularly cytokinins, are present. They observed swelling of the petiole cross-section, especially around the circumference, and death of the central portion after the first week. Plantlet regeneration is usually noticeable within 6 weeks. Nearly 5000 commercially usable plants can be regenerated from a single petiole in 3–4 months.

Using small leaf petiole sections Start and Cummings (4) showed that maximum shoot induction occurred when sections were treated with 0.1 mg/liter NAA, 5.0 mg/liter BA, and 12.5 mg/liter adenine sulfate. They noted that higher concentrations of BA and NAA resulted in callus formation with progressively fewer shoots. Shoot formation appeared first at the base of the veins touching the medium and within weeks was visible on the adaxial side of the leaf. Unrooted shoots were subcultured to a modified MS salt medium lacking growth regulators and containing sucrose at concentrations of 0–30 g/liter. With 0–5 g/liter of sucrose, all shoots died within 14 days; with 12–20 g/liter, normal development occurred; and with 20–30 g/liter, rooting was inhibited. After 6 weeks, the rooted plants were separated from the subcultured clumps and transferred to a soil mixture. After 4 months of greenhouse culture, the rooted plants reached the flowering stage.

Cooke (2) found that small leaf petioles treated with 2.0 mg/liter IAA and 0.08 mg/liter BA produced the highest number of shoots. He noted that shoots began to form after 30 days and multiple shoot production was apparent after 60 days. The number of shoots produced per explant was greater when the abaxial side of the leaf touched the medium. Final cultivation of plantlets was accomplished by aseptic removal, separation, and transfer of individual shoots to a medium containing 30 g/liter sucrose and no growth regulators. After 30–60 days in culture, shoots formed good root systems and grew considerably. Cooke (2) reported that no growth regulators were necessary for rooting and that no inhibition of rooting by sucrose at 30 g/liter occurred as had been observed by Start and Cummings (4).

Torres (5) also evaluated the effects of various NAA and BA concentrations on the initiation of shoots from African violet leaf sections. In his study, NAA and BA in a 4 × 4 factorial experiment were added to an LS medium at 0.0, 0.1, 0.5, and 1.0 mg/liter. He found that shoot production was highest when the medium was supplemented with 0.1 mg/liter NAA and 0.5 mg/liter BA. At higher NAA concentrations, fresh weight decreased and shoot production was inhibited, while at higher BA concentrations, shoot initiation was inhibited.

The effects of leaf size and explant position on the initiation of shoots and roots also was studied by Torres (5), using an LS medium supplemented with 0.1 mg/liter BA and NAA. Each leaf was divided into three vertical sections—base, mid-section, and tip. He concluded that more shoots were obtained when the explant was derived from the base rather than the tip region of the leaf. Also, more shoots were obtained when small leaves were used as the explant source than when large leaves were used.

The purpose of this experiment is first to evaluate the effects of various concentrations of NAA and BA on shoot proliferation and second to study the developmental process of plantlet initiation from the explant. This exercise will allow the reader to learn the techniques for micropropagating African violets as well as the techniques for successfully transferring the shoots to a soil medium. Care should be taken to note the development of shoots from the leaf discs in regard to where on the discs the shoot originates and when the first visible signs of shoot development occur.

Materials Required

1. 90 culture tubes (25 × 150 mm) containing 25 ml of LS culture medium supplemented with hormones according to the following protocol:

   0.0 mg/liter NAA + 0.0 mg/liter BA   30 tubes

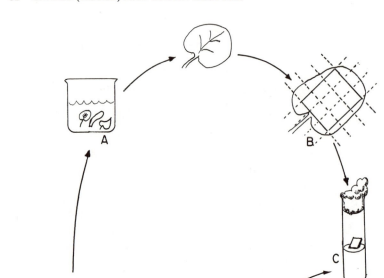

Fig. 6.1.   Steps involved in *in vitro* propagation of African violets.

0.1 mg/liter NAA + 0.1 mg/liter BA	30 tubes
1.0 mg/liter NAA + 1.0 mg/liter BA	30 tubes

2.   10 sterile petri dishes, either glass or plastic (100 mm in diameter)
3.   3 500-ml beakers and one 250-ml beaker
4.   3 pairs of forceps and 2 scalpels
5.   9 plastic racks for holding tubes
6.   Waterproof marking pen and labels
7.   Bunsen or ethanol burner
8.   1000 ml of 20% Clorox solution supplemented with a few drops of Tween-20

9. 1000 ml of sterile distilled water
10. 150 ml of 95% ethanol
11. 12–15 medium to large African violet leaves

Procedures
1. Wipe down all surfaces of the transfer hood as described in Chapter 1. Allow the hood to run for 15 min before beginning transfer operations. Place all the materials listed in the previous section under the hood. Place scalpels and forceps in a 250-ml beaker containing about 100 ml of 95% ethanol.
2. Select 12–15 medium-sized healthy leaves and cut the petiole near the point where it attaches to the stem. Rinse the leaves under running water then transfer the leaves to the 500-ml beakers. Place the beaker under the hood and pour the Clorox solution over the leaves, making certain all leaf surfaces are properly covered (Fig. 6.1A). Leave the leaves in the sterilization solution for 10 min and then pour off the solution. Rinse the leaves three times in sterile distilled water with each rinse lasting approximately 1 min.
3. Place the culture tubes containing each media combination into nine separate racks, as indicated in the protocol. Label each rack with treatment, date experiment began, and cultivar.
4. Transfer each sterilized leaf to a separate sterile petri dish and remove petiole with scalpel. Next, remove the outer edges of each leaf and section it as illustrated in Fig. 6.1B. Once a leaf has been sectioned, weigh each isolated explant and then transfer one leaf section to each culture tube so that the abaxial (underside) side touches the medium (Fig. 6.1C). Each set of 30 tubes should be inoculated with pieces sectioned from the base, mid-section, and tip of the three leaves. Use a different pair of forceps for working with each leaf. Once all culture tubes have been inoculated, place them in racks and incubate in low light at 25°C.

Scheduling

Event	Timing
Isolation of fresh explants	Day 0
First appearance of organogenesis	ca. Day 14 (Fig. 6.1D)
Noticeable shoot formation	ca. Day 30 (Fig. 6.1E)
First subculture	ca. Day 60

Recording Results
1. Record all details of setting up the experiment.
2. Make visual observations at 14-day intervals.
3. Determine fresh weight and shoot number after ca. 60 days.

REFERENCES

1. Bilkey, P. C., B. H. McCown, and A. C. Hilderbrandt. 1978. Micropropagation of African violets from petiole cross-sections. HortScience 12:37–38.
2. Cooke, R. C. 1977. Tissue culture propagation of African violets. HortScience 12(6):549.
3. Gautheret, R. J. 1973. The multiplication of plant tissue cultures. Annu. Rev. Plant Physiol. 6:433–484.
4. Start, N. D. and B. Cummings. 1976. In Vitro propagation of *Saintpaulia ionantha* Wendl. HortScience 11(3):204–206.
5. Torres, K. C. 1982. The effects of various phytohormones on the growth of selected floricultural crops grown *in vitro*. Ph.D. Dissertation, Univ. of Missouri, Columbia.

7

In Vitro Propagation of Lilies

Numerous studies have been reported on the *in vitro* propagation of *Lilium*. Most work has focused on the Asiatic hybrids and the *L. longiflorum* cultivars 'Ace' and 'Nellie White.' Much of this research has investigated the effects of various growth regulators and other culture conditions on the production of various tissues from different explant sources.

For example, Hussey (3) studied the effect of the cytokinin 6-benzyl-amino purine (BA) on the release of axillary buds of *L. pyrenacium* Govan. and an Asiatic hybrid, using adventitious buds from stems as the explant. Optimal production of buds (i.e., 1 to 5 laterals) was promoted by BA concentrations of 2.0–8.0 mg/liter on a Murashige and Skoog (MS) medium. Hussey later found that shoot proliferation in *L. longiflorum* and *L. pyrenacium* was promoted by the use of a cytokinin plus an auxin when pieces of leaves were used as the explant source (4). Stem sections cultured on a MS medium supplemented with 2.0 mg/liter indoleacetic acid (IAA) and 0.5 mg/liter BA also produced approximately 10 bulblets from 1-mm sections.

Stenberg *et al.* (10) obtained optimal bulblet formation from leaf blades of *L. longiflorum* 'Nellie White' when explants were cultured on a MS medium supplemented with 10.0 mg/liter naphthaleneacetic acid (NAA) and 1.0 mg/liter kinetin. This combination, however, was the least effective in promoting shoot formation. Initiation of the bulbs occurred at the base of the explanted leaves, with one to three bulbs forming per explant. Chromosomal counts revealed no chromosomal variation, but histological studies showed that there was a reorganization of embryonic cells at the leaf base into bud primordia as early as 15 days after inoculation. Chen and Holden (1) used two to three young leaves immediately below the flower as their explant source. Plantlets formed from intercalary meristems at the leaf base when the culture medium was supplemented with 5.0 mg/liter NAA and either 0.1, 0.5, or 1.0 mg/liter kinetin.

Hussey (3, 4) reported that optimal production of bulblets from scales of L. longiflorum and L. pyrenacium occurred when a MS medium was supplemented with 2.0–8.0 mg/liter BA. Over 30 lateral buds were produced with 32.0 mg/liter BA, but severe leaf distortion occurred. Hussey also noted that an auxin plus a cytokinin promoted shoot formation. Stimart and Ascher (11) cultured 1-mm thick cross-sections of scales from L. longiflorum 'Ace' on a Linsmaier and Skoog (LS) medium supplemented with various growth regulators. Their data indicated that neither 2iP alone nor 2iP combined with NAA was as effective in bulblet formation as was NAA alone. Combinations of 2,4-D and kinetin or NAA and kinetin were ineffective in plantlet regeneration. Supplementation with 0.03 or 0.3 mg/liter NAA yielded the largest number of bulblets from distal scales, while 0.03 mg/liter NAA was optimal for basal scales. They also noted that NAA alone would induce plantlets from the base of a scale, whereas NAA and 2iP were required to induce plantlets from the tip portion. Larger and more bulbs were obtained when cultures were incubated in the dark, but this occurred at the expense of leaf number and size. Cultures incubated under light exhibited a decrease in bulb number and size but an increase in leaf formation and root number (11). Gupa et al. (2) obtained multiple bulblets from L. longiflorum bulb scale segments and entire scales on a MS medium containing 25 mg/liter adenine sulfate and 1 mg/liter IAA with or without 0.25 mg/liter BA. Callus and/or plantlets have been induced from scales at all times of the year, thus making this system beneficial for the rapid propagation of plantlets (7).

The establishment, growth, and regeneration of plantlets from lily callus cultures without genetic instability has been reported by several workers (1, 3, 7, 8). Sheridan (7) established callus cultures of L. longiflorum by placing the terminal 2-cm stem apices on a LS medium supplemented with 2.0 mg/liter IAA. Cultures without IAA produced buds on the explant, and these formed roots and shoots. Little difference was observed in the growth of callus cultured in the presence or absence of IAA. Kinetin added to the medium in concentrations above 1.0 mg/liter inhibited callus growth, whereas lower concentrations greatly enhanced growth. Prolific callus formation was reported in Asiatic hybrid lilies with 2,4-D as the auxin (4). Simmonds and Cumming (8, 9) reported the propagation of hybrid lilies from callus cultures. They attempted to induce callus from bulb scale segments on a MS medium supplemented with NAA, 2,4-D, or BA alone or in different combinations. Combinations of BA and NAA induced plantlets, and callus was readily induced with 5 μM BA plus 5 μM 2,4-D. Once induced, callus generally differentiated into bulblets which, if not removed, caused a cessation in callus growth. Callus growth was inhibited when the culture medium contained 2,4-D but was maintained on a medium devoid of growth hormones. Reduced levels of ammonium nitrogen also enhanced callus growth. Optimal regeneration was obtained on agar containing a 5.0 μM NAA with exposure to continuous light.

Several studies have been made on the culture of excised *Lilium* embryos *in vitro*. Stimart and Ascher (11) studied the effects of five embryo culture media on the growth of excised embryos from compatible crosses of *L.* × 'Damson' and *L. longiflorum*. Embryos from the intercrosses were excised 35–50 days after pollination. A medium containing all major and minor elements, vitamins, and amino acids was superior in stimulating growth of both immature and mature embryos. Additional growth required transferring the embryos to a medium containing only the macroelements, iron, and a carbohydrate source.

Several comparisons have been made between the use of liquid and semisolid medium for the culture of lily cells. Sheridan (7) reported that cultures responded more favorably when grown in a liquid medium than when grown on a solid medium. Cultures grown on agar occasionally produced plantlets with roots, leaves, and bulbs, regardless of the presence or absence of IAA. Liquid cultures that were frequently subcultured seldom produced plantlets, whereas dense cultures produced a large number of plantlets regardless of the presence or absence of IAA. Simmonds and Cumming (8, 9) found that greater callus growth occurred in suspension cultures than in agar cultures.

The effect of scale section size and position on bulblet formation has been investigated with several *Lilium* species. Robb (6) reported that explants from bulb scale sections of *L. speciosum* Thunb. proliferated and differentiated spontaneously *in vitro*. The regenerative capacity of the explants, however, was seasonal. That is, explants removed in the spring and autumn readily regenerated bulblets (72 and 58%, respectively); whereas scales removed during the summer and winter poorly regenerated bulblets (2 and 0%, respectively). Regeneration also is localized within the scales; sections proximal to the basal plate regenerated more freely than those distal to the basal plate. Regeneration required 15–16 weeks with differentiation originating from the subepidermal region of the scale explant. Stimart and Ascher (11) found that larger and more abundant bulbs were formed from sections derived from the base of the bulb scale compared with sections from the distal section. Maximum production was obtained by using smaller divisions of each scale. They also noted that basal scale explants generated more callus than did distal explants. Gupa et al. (2) noted that the number of bulblets formed and the rate of multiplication was greater when scale segments rather than entire scales were used as the explant source. Further growth of the bulblet, however, was greater with entire scales.

In this experiment the reader will have the opportunity to evaluate the effects of various growth regulators on the initiation of callus and bulb scales and the direct regeneration of shoots from bulb scale segments.

Materials Required

1. 120 culture tubes (25 × 150 mm) containing 25 ml of a culture medium containing: MS basal salts, 1.0 mg/liter thiamin, 100.0 mg/liter *myo*-inositol, 30.0 g/liter sucrose and 8.0 g/liter agar. Supplement the culture medium with the hormones according to the following protocol:

	No. Tubes
0.0 mg/liter NAA + 0.0 mg/liter BA	10
0.1 mg/liter NAA + 0.1 mg/liter BA	10
1.0 mg/liter NAA + 1.0 mg/liter BA	10
10.0 mg/liter NAA + 10.0 mg/liter BA	10

Experiment B	**No. Tubes**
No growth regulators	10
2.5 mg/liter NAA	10
5.0 mg/liter NAA	10
10.0 mg/liter NAA	10
2.5 mg/liter IAA	10
5.0 mg/liter IAA	10
10.0 mg/liter IAA	10
2.5 mg/liter 2,4-D	10
5.0 mg/liter 2,4-D	10
10.0 mg/liter 2,4-D	10

2. 10 sterile petri plates, either glass or plastic (100 mm in diameter)
3. 250-ml beaker and 500-ml beaker
4. 3 pairs of forceps and 2 scalpels with new blades
5. 3 plastic racks for holding tubes
6. Waterproof pen and labels
7. Bunsen or ethanol burner
8. Nylon or wire mesh screen (15 × 15 cm)
9. 250 ml of a 20% Clorox solution supplemented with a few drops of Tween-20
10. 300 ml of sterile distilled water
11. 150 ml of 95% ethanol
12. 1 small to medium-sized lily bulb
13. 2 large healthy *Lilium* bulbs

Procedures

1. Wipe down all surfaces in the transfer hood as described in Chapter 1. Allow the hood to run for 15 min before beginning any transfers. Place all items listed above, except for bulb and 250-ml beaker, in hood.

2. Remove outer layer of damaged scales from the lily bulbs and discard. Under running water, begin removing the rest of the scales making certain to remove all soil; then place the scales in a

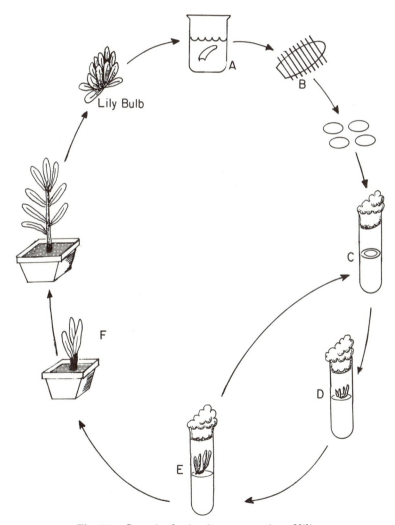

Fig. 7.1.   Steps in the *in vitro* propagation of lilies.

250-ml beaker containing tap water (Fig. 7.1A). Remove the scales down to the point where they become quite narrow and small, then discard the rest of the bulb. Place the nylon mesh over the beaker and place under running tap water so that the scales are gently agitated by the water. Adding a small amount of detergent may be helpful in the cleaning process. Allow the scales to rinse in this manner for 5–10 min. Once the scales have been washed thoroughly, pour off all water, remove the nylon mesh, and transfer the beaker of scales to the laminar flow hood. Pour enough 95% ethanol over the scales to cover them, swirl for 3–4 min, and decant the ethanol. Next pour the Clorox solution over the scales and sterilize for 15 min before decanting. Once the sterilant has been removed, rinse the scales three times with sterile distilled water.

3. Transfer several sterilized scales to a petri dish and section them into 2- to 4-mm thick sections as illustrated in Fig. 7.1B. Note whether each section comes from the base or tip of the scale. After sectioning, weigh each isolated explant and transfer it to one of the culture tubes (Fig. 7.1C). Each set of 10 culture tubes should be inoculated with some sections from the base and tip regions. Once all culture tubes have been inoculated, place them in racks and incubate in low light at 25°C.

Scheduling

Event	Timing
Isolation of fresh explant	Day 0
First appearance of nodules	ca. Day 7–14 (Fig. 7.1D)
Noticeable bulblet formation	ca. Day 30 (Fig. 7.1E)
First subculture	ca. Day 60–90

Recording Results
1. Record all details of setting up the experiment.
2. Make visual observations at 14-day intervals.
3. Determine fresh weight and bulblet number after ca. 60 days.

REFERENCES

1. Chen, C. H. and D. J. Holden. 1975. Cloning *Lilium philadelphiaum* L. by tissue culture. Proc. South Dakota Acad. Sci. 54:143–147.
2. Gupa, P. P., A. K. Sharma, and H. C. Chaturvedi. 1978. Multiplication of *Lilium longiflorum* Thunb. by aseptic culture of bulb scales and their segments. Indian J. Exp. Biol. 16(8):940–942.
3. Hussey, G. 1976. *In vitro* release of axillary shoots from apical dominance in monocotyledonous plantlets. Ann. Bot. 40:1323–1325.

4. Hussey, G. 1977. *In vitro* propagation of some members of the Liliaceae, Iridaceae and Amaryllidaceae. Acta. Hort. 78:303–309.

5. Miller, C. O. and F. Skoog. 1953. Chemical control of bud formation in tobacco stem segments. Am. J. Bot. 40:768–773.

6. Robb, J. M. 1957. The culture of excised tissue from bulb scales of *Lilium speciosum* Thunb. J. Exp. Bot. 8(24):348–352.

7. Sheridan, W. F. 1968. Tissue culture of the monocot *Lilium*. Planta 82:189–192.

8. Simmonds, J. A. and B. G. Cumming. 1976. Propagation of *Lilium* hybrids. I. Dependence of bulblet production on time of scale removal and growth substances. Sci. Horticulturae 5:77–83.

9. Simmonds, J. A. and B. G. Cumming. 1976. Propagation of *Lilium* hybrids. II. Production of plantlets from bulb-scale callus cultures for increased propagation rates. Sci. Horticulturae 5:161–170.

10. Stenberg, N. E., C. H. Chen, and J. G. Ross. 1977. Regeneration of plantlets from leaf cultures of *Lilium longiflorum* Thunb. Proc. South Dakota Acad. Sci. 56:152–158.

11. Stimart, D. J. and D. D. Ascher. 1978. Tissue culture of bulb scale sections for asexual propagation of *Lilium longiflorum* Thunb. J. Am. Soc. Hort. Sci. 103:182–184.

8

In Vitro Propagation of Snake Plant (Sansevieria)

The snake plant, a popular foliage plant, is normally propagated by division or leaf cuttings. Although the plant is easily propagated, both techniques require large amounts of stock material. Also, these methods are slow in terms of the time required to regenerate new plants. Normally, only one plant is produced per leaf cutting, thus placing limitations on the utilization of the propagation material and the number of plants subsequently produced (1, 2).

Leaf tissue has successfully been used as an explant source in many other herbaceous materials. African violets, exacum, flame violets, gloxinia, and others have all been propagated in vitro from leaf tissue. Blazich and Novitzky (1) have described techniques for propagating Sansevieria trifasciata in vitro. Their method involves inducing meristemoids (Step 1) from the leaf surface of Sansevieria, followed by shoot formation (Step 2), and finally rooting (Step 3). In Step 1, explants are placed on a media containing 0.25 mg/liter 2,4-D for 2 weeks. Explants are then transferred to a medium free of hormones for 2 weeks and then transferred again to a medium containing 0.3 mg/liter kinetin until shoots form (Step 2). Shoots greater than 2 cm are excised and rooted in a medium containing 1 mg/liter NAA or rooted under mist in a 1 peat:1 sand (v/v) medium in Step 3.

In this exercise, you will carry out Steps 1 and 2 of Blazich and Novitzky's (1) method for propagating snake plant in vitro. The effects of different cytokinin sources and concentrations on shoot proliferation will be tested. Leaf sections from two Sansevieria cultivars will be used as explants to see if there are any differences between cultivars in the ability to initiate meristemoids and form shoots.

Materials Required
1. 120 culture tubes (25 × 150 mm)
2. 2 scalpels with new blades and 3 pairs of forceps
3. Waterproof pen and stick-on labels

4. Bunsen or ethanol burner
5. Slant racks for holding tubes
6. 500 ml of 95% ethanol
7. 2000 ml of 20% Clorox solution with 0.1% Tween-20
8. Basic medium containing MS salts, 1.0 mg/liter thiamin, 100 mg/liter myo-inositol, 20 g/liter sucrose, and 8 g/liter Difco Bacto-Agar; final pH should be adjusted to 6.3
9. BM-1 medium consisting of basic medium supplemented with 0.25 mg/liter 2,4-D
10. BM-2 medium consisting of basic medium containing no growth hormones
11. BM-3 media consisting of basic medium supplemented with benzyladenine (BA), 6-γ-γ-dimethylallylaminopurine (2iP), or kinetin according to the following protocol:

Cytokinin source	Concentration (mg/liter)
BA	0.0
BA	0.1
BA	1.0
BA	10.0
2iP	0.0
2iP	0.1
2iP	1.0
2iP	10.0
Kinetin	0.0
Kinetin	0.3

12. BM-4 medium consisting of the basic medium plus 1.0 mg/liter NAA.
13. Healthy plants of two Sansevieria cultivars grown under greenhouse conditions.

Procedures
1. Remove leaf sections, 10 cm in length, from the upper half of a 30–40 cm long, actively growing leaf of each cultivar (see Fig. 8.1).
2. Wash the sections in warm soapy water then rinse under running tap water for 15 min. Submerge the tissue in 95% ethanol for 15–30 sec and rinse five times with sterilized distilled water. Next, submerge the leaf tissue in Clorox solution with gentle agitation for 30 min. Finally, rinse the leaf sections three times in sterile distilled water.
3. Cut leaf sections into 1-cm^2 sections.

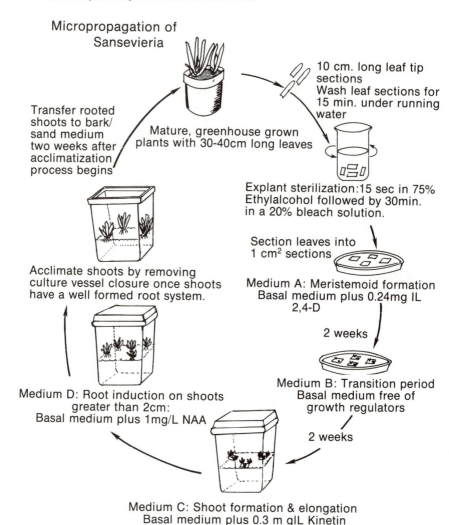

Micropropagation of
Sansevieria

10 cm. long leaf tip
sections
Wash leaf sections for
15 min. under running
water

Transfer rooted
shoots to bark/
sand medium
two weeks after
acclimatization
process begins

Mature, greenhouse grown
plants with 30-40cm long leaves

Explant sterilization:15 sec in 75%
Ethylalcohol followed by 30min.
in a 20% bleach solution.

Section leaves into
1 cm² sections

Acclimate shoots by removing
culture vessel closure once shoots
have a well formed root system.

Medium A: Meristemoid formation
Basal medium plus 0.24mg IL
2,4-D

2 weeks

Medium B: Transition period
Basal medium free of
growth regulators

Medium D: Root induction on shoots
greater than 2cm:
Basal medium plus 1mg/L NAA

2 weeks

Medium C: Shoot formation & elongation
Basal medium plus 0.3 m glL Kinetin

Fig. 8.1. Steps in *in vitro* propagation of snake plants.

4. Dispense 10 ml/tube of BM-1 medium into 40 culture tubes. Inoculate each tube with a 1-cm² leaf section (20 tubes with sections
   from one cultivar and 20 with sections from the other cultivar).
   Incubate cultures for 2 weeks at 25°C under cool white fluorescent
   lights on a 16-hr photoperiod.
5. Dispense 10 ml/tube of BM-2 medium into 40 culture tubes. Two

weeks after initial inoculation, transfer sections to tubes containing BM-2 medium. Incubate as indicated in step 4.

6. Dispense 10 ml/tube of BM-3 media as indicated in protocol in the previous section. You will need 4 tubes for each treatment (2 replications of each treatment for each cultivar). Incubate as indicated in step 4.

7. Approximately 50–75 days after placing the tissue onto BM-3 medium, shoots 20 mm in length can be removed and rooted in a culture medium containing 1.0 mg/liter NAA (BM-4). Rooting will generally occur within 4 weeks; rooted shoots are then transferred to a 1 peat : 1 sand medium and placed under intermittent mist.

Scheduling

Event	Timing
Inoculation of explant onto BM-1 medium	Day 0
Transfer to BM-2 medium without hormones	Day 14
Transfer to BM-3 medium containing cytokinin	Day 28
Transfer to medium for rooting as needed	Day 75–120

Recording Results

1. Record all details of setting up the experiment.
2. Make visual observations at 7-day intervals.
3. Determine the number of shoots produced/explant after 100 days.

REFERENCES

1. Blazich, F. A. and R. T. Novitzky. 1984. *In vitro* propagation of *Sansevieria trifasciata*. HortScience 19(1):122–123.
2. Hartmann, H. T. and D. E. Kester. 1975. Plant Propagation Principles and Practices. 3rd ed. Prentice-Hill, Englewood Cliffs, NJ.

9

Tissue Culture of Strawberry (Fragaria)

Belkengren and Miller (3, 6) were the first to recommend the use of meristem culture for the elimination of viruses from *Fragaria*. Other workers have since used meristem tip culture to eliminate viruses from many commercial strawberry cultivars. Adams (1, 2) was the first to report on the micropropagation of strawberries. He concluded that "it would seem to be possible to obtain an unlimited number of plantlets from a single meristem." Since then, Nishi and Oosawa (7) and Boxus (4, 5) have reported mass propagation of virus-free strawberry plants by *in vitro* culture methods (Figs. 9.1 and 9.2). The techniques described by Nishi and Oosawa (7) involve the use of callus and the regeneration of plantlets from the callus. They found that when strawberry meristems were cultured on a Linsmaier and Skoog medium supplemented with 10^{-5} M BA, 80% of the cultured meristems regenerated plantlets. In the method of Boxus (4), shoots developed from lateral buds formed on the base of each leaflet cultured on a medium containing 1.0 mg/liter BA. He noted that the number of buds formed in this manner was unlimited.

Commercial multiplication of strawberry plants *in vitro* has been described by Boxus (4, 5). The most common procedure utilizes plantlets isolated from meristems, stored on a basal medium, and virus-tested. These plants are rejuvenated by suppressing all the old leaves and roots then transferring them aseptically to fresh basal medium supplemented with 1 mg/liter BA. After 3–4 weeks on this medium, 0 to 3 axillary buds will appear on the lower part of the petiole of the oldest leaf. Within 6–8 weeks, the initial plantlet will appear as a mass of more or less developed buds. Generally no roots or calli are present. The buds can then be transplanted onto a fresh medium. If they are subcultured onto a medium containing a cytokinin, the axillary buds will continue to proliferate, but if they are transplanted onto a medium devoid of cytokinins, the development of axillary buds ceases and nor-

Fig. 9.1. Stages of strawberry micropropagation. Shoot tip explant is isolated from a runner (upper left). The explant develops into a small plantlet (upper right A), which if left on the multiplication medium forms numerous plantlets (upper right B). *In vitro*-derived plantlet (lower left A) is placed on a rooting medium; development after 30 days is shown (lower left B). After being hardened off, plantlets are transferred to *in vivo* conditions (lower right).

mal young plants with roots develop within 4–6 weeks. All of the 74 cultivars tested by Boxus (4) have reacted in the same manner. By this process, it is possible to obtain several million plants from a single mother plant within 1 year.

The purpose of this exercise is to demonstrate the various stages of strawberry micropropagation and evaluate the effects of various NAA, BA, and thiamin concentrations on callus growth and shoot initiation. Shoot will be regenerated from both shoot tips and callus cultures then subsequently rooted, acclimatized, and planted.

Materials Required
1. 500-ml beaker and 250-ml beaker
2. 20 sterile plastic or glass petri plates (15 × 100 mm)
3. Bunsen or alcohol burner
4. 3 pairs of forceps and 3 scalpels

Fig. 9.2.   Stages of plantlet development from strawberry callus.

5.   Waterproof marking pen and labels
6.   Culture tubes (25 × 150 mm) with closures and slant racks for holding them
7.   Stereo or dissecting microscope
8.   500 ml of 10% Clorox solution supplemented with Tween-20 (0.1%)
9.   500 ml of 5% Clorox solution supplemented with Tween-20 (0.1%)
10.   200 ml of 95% ethanol
11.   1000 ml of sterile distilled water

12. Shoot induction medium containing MS micronutrients, 20 g/liter sucrose, 8 g/liter agar, and the following:

Component	Concentration (mg/liter)
CaNO$_3$	1000
KH$_2$PO$_4$	250
KNO$_3$	250
MgSO$_4$	250
Myo-inositol	100
Glycine	2
Nicotinic Acid	0.5
Pyridoxine	0.5
Thiamin	1.0
BA	1.0
IBA	1.0
GA$_3$	0.1

13. Callus maintenance medium containing MS salts, 100 mg/liter *myo*-inositol, 20 g/liter sucrose, and 8 g/liter agar
14. Rooting medium containing MS salts, 20 g/liter sucrose, 8 g/liter agar, and the following:

Component	Concentration (mg/liter)
Myo-inositol	100.0
Thiamin	0.1
Glycine	2.0
Nicotinic acid	0.5
Pyridoxine	0.5
IBA	1.0

15. Soil mixture containing 1 part jiffy mix, 1 part pine bark, and 1 part garden soil by volume
16. 12 healthy strawberry plants

Shoot Initiation
1. Prepare shoot induction medium as described above. Dispense 10 ml/tube into 10 culture tubes and sterilize.
2. Collect shoot tips from either runners or mature plants to use as explants. Under aseptic conditions, peel the shoot tips, leaving only a couple of immature leaves enveloping the meristematic tip. Dip the peeled tips in 10% Clorox solution for 10 min then rinse twice with sterile distilled water. Further peel shoot tips, leaving only the meristematic point. Resterilize for 5 min in 5% Clorox solution. See Fig. 9.1 for illustration of isolation procedure. Utili-

zation of a stereomicroscope may facilitate the removal of immature leaves which envelope the apical meristem.

3. Inoculate each tube with one shoot tip, seal with Parafilm, and
   incubate at 25°C under a 16-hr photoperiod.
4. Transfer on a monthly basis to the same medium to obtain increased growth.

Callus Maintenance

1. Prepare the callus maintenance medium as described above. Divide the medium into four equal aliquots and add NAA, BA, and
   thiamin according to the treatment protocol given in Table 9.1.
2. Dispense 10 ml/tube of medium corresponding to treatments with
   three replications per treatment, and sterilize.
3. Divide callus that has formed from shoot tips in initiation step
   into pea-sized sections and transfer one piece to each culture tube
   prepared in the previous step. Incubate at 25°C.

Rooting of Plantlets

1. Prepare rooting medium as described above and dispense 10 ml/
   culture tube into 20 tubes.
2. Excise shoots that have reached a desirable size (20 mm) in the
   callus maintenance stage (treatment B). Transfer shoots to rooting
   medium and incubate at 25°C. Profuse rooting should occur
   within 1–2 months (Fig. 9.2). Remove closures from cultures once
   plantlets become well rooted.
3. After rooting has occurred, place cultures in well-lighted area at
   25°C for 1–2 weeks to harden off the plantlets.
4. Prepare soil mixture as described above. Place moistened mixture
   into peat pots and transfer hardened shoots to peat pots. Water
   plantlets and spray with Benlate at recommended rates. Place
   potted plants in a covered plastic box to maintain a high relative
   humidity. Gradually open the plastic boxes during the next 1–2
   weeks.

Table 9.1.   Treatment Protocol for Adding NAA, BA, and Thiamin
to Aliquots

Treatment	NAA (mg/l)	BA (mg/l)	Thiamin (mg/l)
A_1	0.0	0.0	0.1
A_2	0.0	0.0	1.0
B_1	0.1	0.1	0.1
B_2	0.1	0.1	1.0
C_1	1.0	1.0	0.1
C_2	1.0	1.0	1.0

5. Once plants exhibit signs of new growth, they may be transferred to the greenhouse to enhance further growth.

Scheduling

Event	Timing
Isolation of meristems	Day 0
Transfer of meristems and removal of any callus	Day 30, 60, 90, 120, etc.
Transfer of plantlets to rooting medium	Day 60, 90, 120, 150, etc.
Hardening of plantlets	1–2 months after transfer to rooting medium
Transfer of plantlets to soil mixture	1–2 weeks after plants are hardened

Recording Results

1. Record the number of meristems inoculated, the number which form shoots and the number which form callus. Next, calculate the percentage of cultures initiating callus and the percentage initiating shoots.
2. Record fresh weights and shoot number initiated from the callus cultures. Record total shoot number and the number of shoots greater than 20 mm. Determine the percentage of calli forming shoots.
3. Record the percentage of shoot number which initiated roots when placed on a root induction medium.

REFERENCES

1. Adams, A. N. 1972. An improved medium for strawberry meristem culture. J. Hort. Sci. 47:263–264.
2. Adams, A. N. 1972. Meristem culture, an extra insurance against viruses. The Grower 26(Feb.):515.
3. Belkengren, R. O. and P. W. Miller. 1962. Culture of apical meristems of *Fragaria vesca* strawberry plants as a method of excluding latent A virus. Plant Dis. Rep. 46:119–121.
4. Boxus, Ph. 1974. The production of strawberry plants by *in vitro* micropropagation. J. Hort. Sci. 49:209–210.
5. Boxus, Ph. 1977. Large scale propagation of strawberry plants from tissue culture. pp. 130–143. *In* Applied and Fundamental Aspects of Plant Cell, Tissue and Organ Culture. J. Reinert and Y. P. S. Bajaj (editors). Springer-Verlag, New York.
6. Miller, P. W. and R. O. Belkengren. 1963. Elimination of yellow edge, crinkle and vein banding viruses and certain other virus complexes from strawberries by excision and culturing of apical meristems. Plant Dis. Rep. 47:298–300.
7. Nishi S. and K. Oosawa. 1973. Mass production method of virus-free strawberry plants through meristem callus. Japan Agr. Res. Quart. 7:189–194.

10

Micropropagation of *Camellia*

Camellias are a popular flowering woody ornamental. Some species are adapted to many climates and are quite tolerant of shade and/or adverse conditions. An additional attractive quality is their colorful display of flowers in fall and winter. Commercially, *Camellia japonica* and *C. sasanqua* are propagated by cuttings. Those species that do not root as readily, such as *C. reticulata*, are generally propagated by approach grafting. Camellias can also be propagated from seed but do not come true. These traditional methods of propagation are time-consuming, and camellia propagation probably would be more efficient if tissue culture techniques were used.

So far, however, only limited success has been achieved in micropropagation of camellias. Creze and Beauchesne (4) were able to establish meristem tip explants of *C. japonica* in tissue culture after incubating the explants in the dark for 15 days. However, of 1400 meristems cultured, only 6 produced plantlets. These authors also reported that callus and plantlets developed *in vitro* from cotyledons cultured on a modified MS medium containing 0.5–5.0 mg/liter BA. However, no data were reported on multiplication rates. More recently, Samartin *et al.* (5) obtained good shoot proliferation from shoot tip explants. They noted that subculturing of the fresh explants onto new medium after the first 4 weeks of culture was required for successful establishment. Shoot proliferation was best when 1.0 mg/liter BA was added to a modified MS medium. Carlisi and Torres (3) found that fresh nodal sections, 10 mm in length, from juvenile *C. sasanqua* initiated shoots best when cultured on a half-strength MS medium supplemented with 1.0 mg/liter BA. Further shoot proliferation could be achieved by culturing *in vitro*-derived shoots on a half-strength MS medium supplemented with 0.1 mg/liter NAA, 2.0 mg/liter BA, and either 5.0 or 10.0 mg/liter of GA_3 (4). Rooting of the microcuttings was achieved by dipping them in 0.5 g/liter of IBA under aseptic conditions and reculturing them on a half-strength MS medium devoid of growth regulators.

The objectives of this exercise are to demonstrate the techniques required for sterilization of woody landscape explants, initiation of nodal explant cultures, subsequent shoot proliferation, and rooting of *in vitro*-derived shoots.

Materials Required
1. 2000-ml beaker and 250-ml beaker
2. 20 sterile plastic or glass petri dishes
3. Bunsen or alcohol burner
4. 3 pairs of forceps and 3 scalpels
5. Waterproof marking pen and labels
6. Culture tubes (20 × 150 mm) with closures and slant racks to hold them
7. 1000 ml of 20% Clorox solution supplemented with Tween-20 (0.1%)
8. 1000 ml of sterile distilled water
9. 200 ml of 95% ethanol
10. 100 ml of a filter-sterilized IBA solution (0.5 g/liter)
11. Basal medium containing half-strength MS salts, sucrose (30.0 g/liter), and agar (8.0 g/liter)
12. Shoot initiation medium consisting of basal medium supplemented with the following (pH 5.0):

Thiamin	1.0 mg/liter
Myo-inositol	100.0 mg/liter
BA	1.0 mg/liter

13. Shoot multiplication medium consisting of basal medium supplemented with the following (pH 5.0):

Thiamin	1.0 mg/liter
Myo-inositol	100.0 mg/liter
BA	2.0 mg/liter
NAA	0.1 mg/liter
GA_3	5.0 mg/liter

14. Rooting medium consisting of basal medium supplemented with thiamin (1.0 mg/liter) and *myo*-inositol (100.0 mg/liter); pH adjusted to 5.0
15. Soil mixture containing 3 parts bark, 1 part peat, and 1 part vermiculite by volume.
16. Juvenile *C. sasanqua* plants.

Establishment and Initiation of Shoots
1. Prepare shoot initiation medium as described above. Dispense 10 ml/tube into 50 25 × 150 mm culture tubes and sterilize.

2. To obtain juvenile *C. sasanqua* plants if none are available, sow 50 seeds in moistened peat; place the peat and seed in a sealed container and store in the dark in a warm, humid environment. Seedlings may also be obtained by scarifing the hard outer seed coat, sterilizing the seed for 10 min in a 20% Clorox solution, and culturing on the root induction medium described above.
3. Obtain 25 softwood cuttings of juvenile *C. sasanqua*, 1 to 3 nodes in length. Sterilize the nodal cuttings for 1 min in 95% ethanol and then for 10 min in 20% Clorox solution. After sterilization, rinse cuttings three times with sterile distilled water.
4. Aseptically remove all leaves and section the cuttings into 10-mm sections, with each section containing one node. Inoculate each culture tube with one explant.
5. Incubate the cultures for 3 weeks at 25°C in total darkness, then move the cultures to low-light conditions at 25°C for the remainder of the 12-week establishment period.

Shoot Multiplication
1. Prepare the shoot multiplication medium as described above. Dispense 10 ml/tube into culture tubes and sterilize.
2. Aseptically collect *in vitro*-derived shoots, 10 mm in length, from cultures initiated in previous section. Each explant should have at least one node. Transfer one explant to each culture tube. Incubate cultures at 25°C under low-light conditions.
3. After 8 weeks on the multiplication medium, all shoots or nodal sections should be recultured onto fresh multiplication medium.
4. After another 8 weeks, shoots or explants with at least one nodal section should be either rooted or recultured onto the multiplication medium.

Rooting of In Vitro-Derived Shoots
1. Prepare the root medium as described above. Dispense 10 ml/tube into culture tubes and sterilize.
2. Prepare 100 ml of an 0.5 g/liter IBA solution and filter-sterilize it.
3. Aseptically collect *in vitro*-derived shoots that are 20 mm or greater in length from multiplication cultures. Soak the basal 7.5 mm of the microcuttings for 30 min in the IBA solution.
4. Transfer one microcutting to each culture tube containing the rooting medium. Incubate the cultures at 20°C under low-light conditions for 8 weeks.
5. Once a mass of roots has formed on a cutting, remove the culture tube closure and place the culture tube under high-light conditions in a growth chamber to harden the cutting.
6. After 2 weeks in the hardening environment, transfer cuttings to

5-cm² peat pots containing the soil mixture described above. Place pots under intermittent mist.

Scheduling

Event	Timing
Isolation of explants	Day 0
Removal from dark culture	Week 3
Transfer to multiplication medium	Week 12
Subculture to fresh multiplication medium	Week 20
Subculture to fresh multiplication medium or transfer to rooting medium	Week 28 or at 8-week intervals
Hardening of rooted microcutting	8 weeks after transfer to rooting medium
Transfer to soil mixture	2–3 weeks after initiation of hardening process

Recording Results

1. Determine the number and percentage of explants initating shoots and the number of contaminated cultures.
2. Determine the average number of shoots produced per explant when placed on the shoot multiplication medium. Also determine the number of shoots greater than 20 mm in length.
3. Determine the number of roots per shoot when placed on the root initiation medium.

REFERENCES

1. Beauchesne, G. 1979. Tissue culture pot pourri. pp. 7–13. *In* Proceedings of the Int. Camellia Society Congress.
2. Bennett, W. Y. and P. Scherbert. 1982. *In vitro* generation of callus and plantlets from cotyledons of *Camellia japonica*. Camellia J. 37(1):12–15.
3. Carlisi, J. A. and K. C. Torres. 1985. Establishment of cultures from primary explants of *Camellia sasanqua*. In press.
4. Creze, J. and N. G. Beauchesne. 1980. Camellia cultivation *in vitro*. Intern. Camellia Soc. J. 12:31–34.
5. Samartin, A., A. M. Vieitez, and E. Vieitez. 1984. *In vitro* propagation of *Camellia japonica* seedlings. HortScience 19(2):225–226.
6. Torres, K. C. and J. A. Carlisi. 1986. Shoot and root organogenesis of *Camellia sasanqua in vitro*. Plant Cell Reports. In press.

11

Propagation of Fern (*Nephrolepis*) Through Tissue Culture

Ferns are cultivated commercially for their ornamental foliage, and demand for them as an indoor foliage plant has increased dramatically over the last decade. Fern tissue culture has been used as a research tool to study the developmental potentialities of leaf primordia since the early 1950s (8, 17). More applied research directed toward the *in vitro* propagation of ferns was begun in the early 1960s. This work involved the differentiation of *Pteris cretica* L. into gametophytic or sporophytic tissue (3). Knauss (11) described a partial tissue culture method for the propagation of selected ferns from spores. His method involved the germination of spores in a sterile culture and the subsequent periodic sectioning and transfer of gametophytic tissue. This tissue was homogenized in a blender with a half-strength MS salts solution for 5 sec and then inoculated onto culture medium for sporophytic plantlet development. Excised leaf primordia of *Osmunda cinnamomea* (17), *Adiantum pedatum*, and other ferns have also been propagated through tissue culture (16).

The conventional method employed for the propagation of ferns has been by runners. Ferns derived in this manner produce a single growth center from which only a few fronds develop at one time. In contrast, ferns derived through tissue culture techniques produce several growth centers from each of which numerous new fronds are produced (15).

A partial list of fern genera that have been successfully propagated *in vitro* is presented in Table 11.1.

The purpose of this study is to evaluate the techniques required for propagating ferns. This exercise will utilize runners from fern stock plants that will be placed on a multiplication medium. Once the shoots reach a transferable stage, they can then be transferred to a root induction medium.

Table 11.1. Fern Genera That Have Been Successfully Propagated *In Vitro*[a]

Family	Species	Explant Source	Morphogenic Response	Reference
Adiantaceae	*Adiantum cuneatum* Langed	Rhizome tip	Adventitious tip	14
	Adiantum tenerum Swartz	Homogenized gametophytic tissue	Sporophytic plants	11
Blechnaceae	*Woodwardia fimbriata* Grant	Rhizome tip	Adventitious shoots	9, 14
Cyatheaceae	*Alsophlia australis* R. Br.	Rhizome tip	Adventitious shoots	9, 14
Davalliaceae	*Davallia bullata* Wall	Homogenized plants from culture	Sporophytic plants	6
	Nephrolepis cordifolia (L.) Prest	Runner tip	Sporophytic plants	18
	Nephrolepis exaltata (L.) Schottvar. bostoniensis	Runner tip	Adventitious shoots	4, 5, 9, 14, 16
	Nephrolepis falcuta f. furcans Proctor	Runner tip	Adventitious shoots	2
Dennstaedtiaceae	*Nephrolepis pendula*	Runner tip	Adventitious shoots	1
Dyopteridoidea	*Microlepia strigosa* K. Presl.	Rhizome tip	Adventitious shoots	9, 14
Plalycerioideae	*Cyrtomium falcatum* Presl.	Homogenized gametophytic tissue	Sporophyte plants	11
	Platycerium bifurcatum	Homogenized gametophytic tissue	Sporophyte plants	6, 11
	Platycerium stemaria Beauvois	Shoot tip	Adventitious shoots	10
Pteridordeae	*Pteris argyraea* Moore	Rhizome tip	Adventitious shoots	14
	Pteris cretica L.	Leaf stem	Gametophytes, sporophytes	3
	Pteris ensiformis Burum	Homogenized gametophyte tissues	Sporophytic plants	11
	Pteris vittata L.	Rhizome callus	Sporophytes, gametophytes	12
Thelypteudaceae	*Cyclosoues dentatus* (Forsk) Ching	Root apex explants	Gametophytes, sporophytes	13

[a] Assignment of genera to families according to Crabbe et al. (7).

Materials Required
1. 1000-ml beaker and 250-ml beaker
2. 20 sterile glass or plastic petri dishes
3. 3 pairs of forceps and 3 scalpels with new blades
4. Waterproof marking pen and labels
5. Bunsen or alcohol burner
6. 125-ml Erlenmeyer flasks with closures to serve as culture vessels
7. 1000 ml of 20% Clorox solution supplemented with a few drops of Tween-20
8. 1000 ml of sterile distilled water
9. 200 ml of 95% ethanol
10. Culture medium consisting of MS basal salts, sucrose (30.0 g/liter), agar (8.0 g/liter), and the following:

Component	Concentration (mg/liter)
Myo-inositol	100
Thiamin	1
Pyrodoxine HCl	1
Nicotinic acid	1
Sodium phosphate (NaH_2PO_4)	150
NAA	1
2iP	10
Adenine sulfate	30

11. Stock plants of *Nephrolepis*

Procedures
1. Prepare the culture medium as described above. Dispense 50 ml of medium per culture flask, cap, and sterilize.
2. Collect 25–30 nondifferentiated runners, 5–10 cm in length, from stock plants of *Nephrolepis*. Surface-sterilize the runners in 20% Clorox solution for 20 min. Following sterilization, section the runners into 2.5-cm-long sections and place in a sterile 250-ml beaker. Rinse the runners three times with sterile distilled water.
3. Inoculate each culture flask with three runner sections, cap the flasks, and incubate cultures at 25°C under low-light conditions. Growth should appear within 6–8 weeks.
4. Subculture the cultures 8–12 weeks after initial inoculation. To subculture, divide clumps of plantlets into 1-cm^2 sections and transfer to the same medium for further shoot proliferation.
5. Plantlets can be transferred to a suitable soil mixture (3 bark : 1 peat : 1 vermiculite) and rooted under high relative humidity.

Scheduling

Event	Timing
Isolation of explants	Day 0
Transfer of cultures to a multiplication medium	Every 6–8 weeks
Transfer of shoots to a rooting medium	6–8 weeks after previous transfer
Fully rooted plantlets	4–6 weeks after transplanting in soil mix

Recording Results

1. Record the number of transferable shoot 6–8 weeks after placement on a shoot multiplication medium.
2. Record the number of plants which survived after transplanting to a soil mix.

REFERENCES

1. Ault, J. 1985. Personal communication.
2. Beck, M. J. 1980. The effects of kinetin and naphthaleneacetic acid ratios on *in vitro* shoot multiplication and rooting in the fish tail fern. M.S. Thesis, Univ. of Tennessee, Knoxville.
3. Bristow, J. M. 1962. The controlled *in vitro* differentiation of callus derived from a fern *Pteris cretica* L. into gametophytic and sporophytic tissues. Dev. Biol. 4:361.
4. Burr, R. W. 1975. Mass propagation of the Boston fern through tissue culture. Proc. Int. Plant Prop. Soc. 25:122.
5. Burr, R. W. 1976. Mass propagation of ferns through tissue culture. *In Vitro* 12:309.
6. Cooke, R. C. 1979. Homogenization as an aide in tissue culture propagation of *Platycerium* and *Dovalia*. HortScience 14:21.
7. Crabbe, J. A., A. C. Jermy, and J. M. Mickel. 1975. A new arrangement for the pteridophyte herbarium. Fern Grz. 11:141.
8. Culter, E. G. and C. W. Warddaw. 1963. Induction of buds on older leaf primordia in ferns. Nature 173:440–441.
9. Harper, K. L. 1976. Asexual multiplication of *Leptosporangiate* ferns through tissue culture. M.S. Thesis, Univ. of California-Riverside.
10. Hennen, G. R. and T. J. Sheehan. 1978. *In vitro* propagation of *Platycerium stemaria* (Beauvois) Desv. HortScience 13:245.
11. Knauss, J. 1976. A partial tissue culture method for pathogen-free propagation of selected ferns from spores. Proc. Fla. State Hortic. Soc. 89:363.
12. Kshiroagar, M. K. and A. R. Mehta. 1979. *In vitro* studies in ferns: growth and differentiation in rhizome callus of *Peteris vittata*. Phytomorphology 28:50.
13. Mehra, R. N. and H. K. Plata. 1971. *In vitro* controlled differentiation of the root callus of *Cyclosorus dentatus*. Phytomorphology 21:367.
14. Murashige, T. 1974. Plant propagation through tissue cultures. Annu. Rev. Plant Physiol. 25:135.

15. Murashige, T. 1977. Clonal crops through tissue culture. pp. 392–403. *In* Plant Tissue Culture and its Bio-Technological Application. W. Barz, E. Reinhard, and M. H. Zenk (editors). Springer-Verlag, New York.
16. Padhya, M. A. and A. R. Mehta. 1982. Propagation of fern (*Nephrolepis*) through tissue culture. Plant Cell Rep. 1:261–263.
17. Steeves, T. A. 1961. A study of the developmental potentialities of excised leaf primordia in sterile culture. Phytomorphology 11:346–358.
18. Sulklyan, D. S. and P. N. Mehra. 1977. *In vitro* morphogenetic studies in *Nephrolepis cordifolia*. Phytomorphology 27:396.

12

Establishment and Maintenance of Carrot Callus

Carrot [*Daucus carota* L. subsp. *sativus* (Hoffm.)] is a cool season plant grown for its edible storage tap root. In 1983, the carrot ranked seventh among vegetable crops produced in the United States with a market value of over $100 million. The carrot is derived from the wild carrot (*D. carota* L. subsp. *carota*) or Queen Anne's Lace and belongs to the Umbelliferae family.

The first cultures from nontumorous plant cells were from roots of carrot (3, 4). The cultures were initiated and maintained on a Knops medium supplemented with Berthelot's mixture of salts, glucose, thiamin, cysteine hydrochloride, gelatin, and IAA. Carrot callus could be maintained on this medium indefinitely without any apparent reduction in growth. Nobercourt (4) noted that roots formed on the carrot callus in his experiments. When this callus was transferred to a liquid medium, the roots would continue to develop.

Early work on carrot morphogenesis was conducted by Steward (7) and Reinert (5) working with suspension and callus cultures, respectively. Steward (7) reported that carrot cell suspensions cultured in the presence of CM resulted in the formation of multicellular masses which if plated onto a semisolid medium, would form meristematic nodules resembling carrot embryos and which would eventually develop into buds and shoots. If the multicellular masses were left in suspension, root initiation would occur but shoot proliferation would be inhibited.

Reinert (5) reported that callus cultured on a complex medium described by Reinert and White (6) for several months would gradually change in texture with the outer callus surface covered with nodules that under microscopic observation appeared to be normal bipolar embryos. Callus grown on a medium containing 7% CM and 57 μM IAA which would initiate roots if then transferred to a hormone-free medium or if left on the auxin-containing medium for a long period (5).

Since the discovery of somatic embryogenesis in carrot culture was made by Steward and Reinert, it has become a model system for investigating many biochemical, physiological, and genetic aspects of plant cell culture. Extensive work by Ammirato and his co-workers (1, 2) has led to the development of highly productive schemes for plant production from somatic embryos of carrot.

Materials Required
1. 10 glass or plastic petri dishes (100 mm in diameter)
2. 2 pairs of forceps and 2 scalpels
3. Sterile #2 cork borer and glass rod that fits it
4. 1000-ml beaker and 250-ml beaker
5. Waterproof marking pen and labels
6. 4 plastic slant racks to hold solid culture tubes
7. Analytical balance
8. Bunsen or ethanol burner
9. 500 ml of 20% Clorox solution supplemented with a few drops of Tween-20
10. 500 ml of sterile distilled water
11. 200 ml of 95% ethanol
12. Solid B5 medium supplemented with the following:

0.1 mg/liter 2,4-D + 0.1 mg/liter BA	10 tubes
0.1 mg/liter 2,4-D + 1.0 mg/liter BA	10 tubes
1.0 mg/liter 2,4-D + 0.1 mg/liter BA	10 tubes
1.0 mg/liter 2,4-D + 1.0 mg/liter BA	10 tubes

13. Solid B5 medium supplemented with the following:

0.1 mg/liter 2,4-D + 0.1 mg/liter kinetin	10 tubes
0.1 mg/liter 2,4-D + 1.0 mg/liter kinetin	10 tubes
1.0 mg/liter 2,4-D + 0.1 mg/liter kinetin	10 tubes
1.0 mg/liter 2,4-D + 1.0 mg/liter kinetin	10 tubes

14. 1 or 2 healthy, undamaged, and regularly shaped carrot roots.

Procedures
1. Wipe down and turn on laminar flow hood 15 min before doing any work in the hood. Flame-sterilize instruments.
2. Clean carrot root by scrubbing under running tap water to remove any surface soil. Trim the carrot into 100-mm sections and place them in a 1000-ml beaker. Cover with 20% Clorox solution for approximately 20 min and then decant the Clorox solution. Rinse the explant three times in sterile distilled water covering the tissue with each rinse.
3. While carrot sections are being sterilized, dispense media into culture tubes according to protocols in the previous section. Place

Carrot Micropropagation
and Somatic Embryogenesis

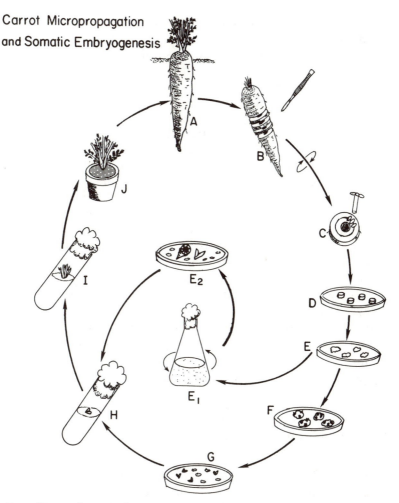

Fig. 12.1. Stages of carrot micropropagation and somatic embryogenesis. (A) Mature carrot plant with tap root. (B) After sterilizing the root in 20% Clorox for 20 min, slice the root into 1–2-mm-thick sections. (C) Using a cork borer, punch out cylinders containing part of the cambial region. (D) Plate carrot cylinders which have been cut 1–3 mm in length onto a medium supplemented with 2,4-D for callus initiation. (E) Callus can be recultured onto an identical medium, onto a callus embryogenic medium (F), or into a cell suspension medium containing 4.5 μM 2,4-D (E$_1$). Somatic embryos can be induced by either resuspending the cell suspension in a medium devoid of 2,4-D or plating the calli forming embryos onto a medium devoid of 2,4-D(G). The somatic embryos can then be removed and plated out and allowed to develop into whole plants (E$_2$–J).

tubes in racks and label them with respective treatments, medium, and date.
4. Transfer sterilized carrot slices to a sterile petri dish. Using a sterile cork borer, punch out 8–10 cylinders of tissue from the secondary phloem/cambial region of the carrot slices (Fig. 12.1). *Do not* punch out more than two cylinders of tissue at one time, as you may have difficulty in removing them from the cork borer. Using a sterile glass rod, push the cylinders from the cork borer into a sterile petri dish. Using a sharp, sterile scalpel, remove the ends of the tissue cylinder. Next cut the remaining portion of the cylinder into 2- to 3-mm-thick sections. Using a sterile petri plate on the balance, weigh each explant section individually and inoculate into the culture tubes, placing one explant per tube. Repeat this procedure until all culture tubes have been inoculated. Use a different forcep for each explant and flame the forceps between transfers. Flame the lid of the culture tubes immediately after opening and before closing.

Callus should be removed from the primary explant after 45 days. The calli can be subcultured onto an identical medium for further callus growth or subcultured into a liquid medium containing 2,4-D to initiate a cell suspension culture (Fig. 12.1).

Scheduling

Event	Timing
Isolation of explants	Day 0
First subculture	Day 60
Isolation of callus	Day 120

Recording Results
1. Record all details of setting up the experiments.
2. Make visual observations at 14-day intervals.
3. Determine fresh weight gain at 60 and 120 days.

REFERENCES

1. Ammirato, P. V. 1984. Induction, maintenance and manipulation of development in embryogenic cell suspension cultures. pp. 139–151. *In* Cell Culture and Somatic Cell Genetics of Plants, Vol. 1, I. K. Vasil, ed. Academic Press, New York.
2. Ammirato, P. V. 1986. Carrot, pp. 457–499. *In* Handbook of Plant Cell Culture. Techniques and Applications. D. A. Evans, W. R. Sharp, and P. V. Ammirato (editors). Macmillian Publishing, New York.
3. Gautheret, R. J. 1939. Sur la possibilité de realiser la culture indefinie des tissus de tubercules de Carotte. Compt. Rened. 208:118–121.

4. Nobercourt, P. 1939. Sur les radicelles naissant des cultures de tissus du tubercule de Carotte. Compt. Rend. Soc. Biol. 13:1271–1272.

5. Reinert, J. 1959. Uber die Kontrolle der Morphogenese und die Induktion von Adventive embryonen an Gewebekulturen aus Karotten. Planta 53:318–333.

6. Reinert, J. and P. R. White. 1956. The cultivation in vitro of tumor tissues and normal tissue of *Picea glauca*. Planta 53:18–24.

7. Steward, F. C., M. O. Mapes and K. Mears. 1958. Growth and organized development of cultured cells. Am. J. Bot. 45:705–708.

13

Callus Induction in Grasses

The use of tissue culture techniques in the study of growth, metabolism, and differentiation has been widely developed for the dicotyledons, but the monocots, especially the grasses, have received little attention. One reason for this was that until recently the establishment of *in vitro* cultures of monocots was quite difficult. However, this problem was overcome with the discovery that a large concentration of auxin, in comparison to what is used with dicots, would easily induce callus initiation.

CALLUS INITIATION

The most commonly used and generally most effective auxin for inducing and maintaining callus in the cereals is 2-4D; IAA also is frequently used. In the case of IAA, 50–100 mg/liter is required for callus induction in wheat and rice (9). This high requirement may indicate that more of the IAA is converted into an ineffective form in the monocotyledons than in the dicotyledons. Yamada (9) found that both rice and wheat callus formed at high auxin concentrations regardless of kinetin concentration, indicating that auxin and not kinetin is absolutely essential for callus formation. It was also shown that in these species when IAA was used in the same concentration range as 2,4-D, the 2,4-D bound with proteins to form a 2,4-D/lysine-rich histone complexes during the early stages of callus initiation. Shimada *et al.* (8) also found that kinetin had no effect on callus initiation of wheat. Optimal callus was formed with 2,4-D at 1.0–10.0 mg/liter or IAA at 50.0 mg/liter. The best subsequent callus growth was obtained when 1.0 g/liter of casein hydrolysate or 1% coconut milk was used. Vigorous growth was also obtained with 0.5–2.0 mg/liter 2,4-D.

Carew and Schwarting (2) found that callus induction of rye (*Secale cereale*) was optimal when dissected embryos were placed on a media of Hellers salts, 2% sucrose, 0.5% yeast extract, and 1.0 mg/liter 2,4-D. The subculture callus grew optimally on the same media supplemented with casein hydrolysate. Lowe and Conger (6) found that callus of tall fescue (Festuca arundinacea) initiated from mature embryos on a MS media supplemented with 9.0 mg/liter 2,4-D. After transfer, the callus was maintained on the same media containing 5.0 mg/liter 2,4-D. Chen et al. (3) induced callus of big bluestem (Andropogon gerardii Vitman) from young inflorescences on a LS medium supplemented with 5.0 mg/liter 2,4-D and 0.2 mg/liter kinetin when grown in the dark at 25°C. The callus was maintained by subculturing onto the same medium containing 5.0 mg/liter 2,4-D. Conger and Carabia (4) induced callus formation of orchardgrass (Dactylis glomerata L. 'Boone') on a modified Schenk-Hilderbrandt medium containing 15.0 mg/liter 2,4-D and 2.15 mg/liter kinetin. The callus was maintained by subculturing at 3-week intervals on a medium containing 5.0 mg/liter 2,4-D. Kasperbauer et al. (5) established cultures of an annual ryegrass (Lolium multiflorum Lam) × tall fescue (Festuca arundinacea Schreb.) cross on a MS medium at levels of 2,4-D ranging from 2.0 to 4.0 mg/liter. The cultures were then maintained on medium containing 2.0 mg/liter 2,4-D.

Atkin and Barton (1) established cultures of 12 grasses by incubating detached roots or whole seedlings on a LS medium supplemented with 1.0 mg/liter 2,4-D. The callus was shown to develop from the pericycle of the roots and from the embryos of the seeds. Maintenance of calli was obtained on the same medium. The addition of kinetin usually reduced the production of callus regardless of the auxin source. Although both 2,4-D and IAA promoted the formation of callus, 2,4-D was clearly more effective than IAA. Also, 2,4-D had a further advantage in that the callus did not brown as quickly as it did when IAA was used. The callus, whether isolated from roots or embryos, was morphologically indistinguishable; however, different species may vary in color and texture.

Reynolds and Murashige (7) used a 2,4-D concentration of 100.0 mg/liter to initiate callus from palms. The MS medium used was supplemented with 170 mg/liter of additional phosphate, 30 g/liter sucrose, 40 mg/liter adenine sulfate, 100 mg/liter myo-inositol, and 0.4 mg/liter thiamin.

PLANT REGENERATION

To achieve embryogenesis in palms, Reynolds and Murashige (7) removed the cytokinin and auxin from the medium. After 6 weeks, small rootlike protrusions could be seen and the calli were transferred to the same media. After another 4 weeks, the rootlike structures developed into small seedlings.

Kasperbauer et al. (5) obtained regeneration from callus of F_1 hybrids of annual ryegrass × tall fescue by reducing the 2,4-D in the medium to 0.25 mg/liter. Cultures that had been subcultured at 3- to 4-week intervals were induced to regenerate plantlets containing the same chromosome number as and identical in appearance to the source plant. However, cultures that had been subcultured at 8-week intervals contained a double chromosome number.

Conger and Carabia (4) obtained organogenesis of orchardgrass when the 2,4-D concentration of their medium was reduced to below 1.0 mg/liter. Callus had been initiated on nearly every explant used, but the initiation of shoots was less prolific. Lowe and Conger (6) used a medium containing 0.5 mg/liter of 2,4-D to test for root and shoot formation in tall fescue. They tested organogenesis at three different periods, 45 days apart, to determine if repeated subculture would affect the tissues' regenerative ability. They found that regardless of subculture, root formation was about 45%. The number of calli forming shoots, however, decreased with each subsequent subculture, ranging from 9.6% with calli from the first subculture to 3.1% with calli from the third subculture.

Chen et al. (3) found that numerous plants regenerated from the callus of big bluestem when the concentration of 2,4-D was reduced to below 2.0 mg/liter. With ryegrass, calli differentiated root and shoot primordia on a MS medium supplemented with 1.5 mg/liter 2,4-D, 6.5 mg/liter IAA, and 2.15 mg/liter kinetin. They noted that the addition of 1 ml coconut milk with or without 2 mg/liter IAA and 2 mg/liter zeatin stimulated chlorophyll synthesis in the callus. Subsequent subcultures grown on half-strength MS medium plus 0.75 mg/liter 2,4-D, 3.25 mg/liter IAA, and 1.075 mg/liter kinetin resulted in the formation of numerous normal and a few albino plantlets. They observed that these calli maintained for 18 months through eight subcultures retained their totipotent capacity.

As noted, the induction of callus in the monocots requires a larger than normal application of auxin, usually 2,4-D. For regeneration to occur, the 2,4-D concentration must be decreased or the 2,4-D replaced with a weaker auxin source such as IAA or NAA. It appears that the auxin component is the most critical for regeneration in monocots, as most species have been induced to regenerate without the addition of cytokinin when the auxin level is lowered. The typical sequence of callus initiation, growth, and regeneration in grasses is illustrated in Fig. 13.1.

This experiment will demonstrate the techniques required for the

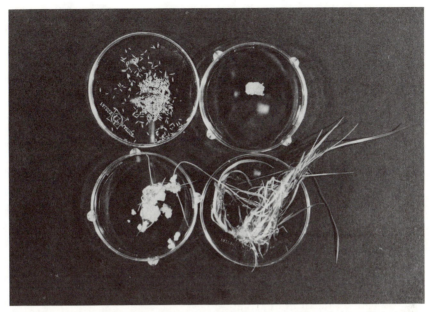

Fig. 13.1. Stages of grass development *in vitro*. Caryopses are used as explants (upper left), which develop into a small callus (upper right). When calli are placed on a regeneration medium, small shoots begin to develop (lower left) and these eventually lead to plantlet formation (lower right).

initiation of callus from grass caryopses (seed), the subsequent subculture of the callus followed by the regeneration of plantlets from callus.

Materials Required
1. 2 250-ml beakers and a 1000-ml beaker
2. 2 wire inoculating loops
3. Timer
4. 40-position small stainless steel culture rack
5. Waterproof marking pen and 4 labeling tags with ties
6. Bunsen or ethanol burner
7. 250 ml of 95% ethanol
8. 250 ml of 20% Clorox solution
9. 300 ml of sterile distilled water
10. 40 culture tubes (20 × 150 mm) containing 15 ml of LS culture medium and the following hormones:

0.0 mg/liter 2,4-D + 1.0 mg/liter BA	10 tubes
0.1 mg/liter 2,4-D + 1.0 mg/liter BA	10 tubes
1.0 mg/liter 2,4-D + 1.0 mg/liter BA	10 tubes
10.0 mg/liter 2,4-D + 1.0 mg/liter BA	10 tubes

11. 500 *Agrostis palustris* 'Penncross' creeping bentgrass caryopses (seed)

Procedures
1. Wipe down all surfaces of the transfer hood. Allow the hood to run for 15 min before beginning transfers.
2. Prepare media and dispense into culture tubes as described in protocol in the previous section. Label with treatment, date, and name.
3. Place bentgrass seed in a 250-ml beaker. Pour 50–100 ml ethanol over the top of the caryopses and sterilize under the hood for 5 min. Pour off the ethanol, pour the 20% Clorox solution over the seed, and sterilize the cryopses in this solution for 15 min; decant the sterilant. Following sterilization, rinse the caryopses three times in sterile distilled water, using approximately 100 ml of sterile water per rinse.
4. Flame-sterilize the inoculating loop and allow it to cool. Using the inoculating loop, place 1 to 3 caryopses into each culture tube until all have been inoculated. Place the tubes in the culture racks and incubate in the dark at 25°C.
5. Callus should be subcultured at 60-day intervals. Callus is transferred by taking a callus colony and dividing into a small pealike segment then transferring to a fresh medium. Callus can be transferred to a callus maintenance medium for further proliferation or may be transferred to a shoot induction medium. Shoots can be removed from callus once the shoot reaches a height of at least 25 mm. These shoots can be transferred to a root enduction medium for root initiation. After 30–45 days, or when a substantial number of roots have formed, the shoots can be acclimatized by removing the closure from the culture tube and incubating under high light conditions (light bank) for 1–2 weeks. The plantlet can then be transferred to a soil mix and placed under mist in a greenhouse.

Scheduling

Event	Timing
Isolation of fresh explant	Day 0
Germination of seed	Day 3–7
Noticeable callus formation	ca. Day 30
First subculture	ca. Day 60–90

Recording Results
1. Record all details of setting up the study.
2. Make visual observations at 14-day intervals.
3. Determine callus fresh weight after ca. 60 days.

REFERENCES

1. Atkin, R. K. and G. E. Barton. 1973. The establishment of tissue culture of temperate grasses. J. Exp. Bot. 21:689–699.
2. Carew, D. P. and A. E. Schwarting. 1958. Production of rye embryo callus. Bot. Gaz. 119:237–239.
3. Chen, C. H., N. E. Stenberg, and J. G. Ross. 1977. Clonal propagation of big bluestem by tissue culture. Crop Sci. 17:847–850.
4. Conger, V. B. and J. V. Carabia. 1978. Callus induction and plantlet regeneration in orchardgrass. Crop Sci. 18:157–159.
5. Kasperbauer, M. J., R. C. Buckner, and L. P. Bush. 1979. Tissue culture of annual ryegrass × tall fescue F_1 hybrids. Callus establishment and plant regeneration. Crop Sci. 19:457–460.
6. Lowe, K. and B. V. Conger. 1979. Root and shoot formation from callus cultures of tall fescue. Crop Sci. 19:397–400.
7. Reynolds, J. F. and T. Murashige. 1979. Asexual embryogenesis in callus cultures of palms. *In Vitro* 15:383–387.
8. Shimada, T., T. Sasakuma, and K. Tsunewaki. 1969. Tissue culture of wheat tissues. Callus formation, organ redifferentiation and single cell culture. Can. J. Genet. Cytol. 11:294–304.
9. Yamada, Y. 1977. Tissue culture studies on cereals. *In* Plant Cell, Tissue and Organ Culture. J. Reinert and Y. P. S. Majaj (editors). Springer-Verlag, New York.

14

Initiation of Adventitious Shoots of *Begonia* × *hiemalis*

The high degree of organogenic potential exhibited by *Begonia* × *hiemalis* explants has been used to develop micropropagation procedures for this species. Micropropagation of *Begonia* takes advantage of the induction of a very large number of adventitious buds or shoots in response to various hormones. Several studies have indicated the potential for producing vast numbers of plants by these means (2, 3, 5, 9); however, in practice this potential is rarely realized because many of these adventitious buds remain too small to be rooted (5). Considerable plant loss may occur at the rooting and acclimatization stages.

Explants can be taken from the leaf petiole (2, 5, 10, 11, 12), flower peduncle (1), or leaf lamina (7). The growth conditions of the stock plants may affect the *in vitro* initiation of shoots and roots (1, 10). Likewise, the types and combinations of growth regulators in the growth medium as well as variation among cultivars may influence organ formation (2, 5, 10, 11).

Welander (10, 11) evaluated root and shoot formation from petiole explants in 17 cultivars of *B.* × *hiemalis* and found that shoot formation was favored with a low ratio of NAA:BA while a high ratio of NAA:BA favored root formation. Mikkelsen and Sink (5) noted that NAA at 0.1 mg/liter and BA at 0.4 mg/liter resulted in the greatest number of adventitious shoots from *B.* × *hiemalis* 'Schwabenland Red.' From each 5-mm petiole section, approximately 45 shoots were produced.

Shoot formation from *B.* × *hiemalis* is also influenced by culture temperature as well as by the temperature at which the stock plants are grown. Fewer shoots are produced per explant at higher temperatures, e.g., 25°C, than at lower temperatures, e.g., 15°–20°C (1, 2, 3). In this chapter details on the micropropagation of *Begonia* will be discussed.

Materials Required
1. 10 sterile glass or plastic petri dishes (100 mm in diameter)
2. 3 pairs of forceps and 3 scalpels

3. Waterproof marking pen
4. 2000-ml plastic beaker and 250-ml glass beaker
5. Analytical balance
6. Bunsen or ethanol burner
7. Culture tubes (25 × 150 mm) and plastic slant racks to hold them
8. 1000 ml of 20% Clorox solution supplemented with a few drops of Tween-20
9. 500 ml sterile distilled water
10. 200 ml 95% ethanol
11. Basic culture medium consisting of the following components:

Shoot induction medium	**Amount**	
MS salts[1]		
Sucrose	10.0	g/liter
Agar	8.0	g/l
NAA	0.5	mg/liter
Myo-inositol	100	mg/liter
Thiamin	1.0	mg/liter

Root induction medium	**Amount**	
MS basal salts[1]		
Sucrose	20.0	g/liter
Agar	8.0	g/liter
IBA	2.0	mg/liter
Thiamin	1.0	mg/liter
Myo-inositol	100.0	mg/liter

12. *Begonia* × *hiemalis* plants

Procedures

1. Wipe down and turn on the laminar flow hood 15 min before doing any work under the hood. Place all instruments in a 250-ml beaker filled with at least 125 ml of ethanol; flame-sterilize them before use.
2. Prepare basic shoot induction medium as described above and supplement it according to the following protocol:

Treatment A		**Treatment B**	
BA	10 mg/liter	BA	1.0 mg/liter
		Nicotinic acid	1.5 mg/liter
		Pyridoxine HCl	1.5 mg/liter

[1] Prepared as described in Chapter 2 or from prepackaged salts.

Dispense 25 ml/tube of supplemented medium into tubes, with 10 replications per treatment, and sterilize.

3. Select 12 healthy *Begonia* × *hiemalis* leaves, 5–10 cm in diameter. Surface-sterilize leaves by immersion in the Clorox solution for 5 min then rinse three times in sterile distilled water.

4. Place leaves in petri plates and cut into 1.5-cm² sections; discard the outer leaf margin. Each leaf section should contain part of the midrib or a lateral vein.

5. Excise petioles into 5-mm sections and discard portion of the petiole farthest from the leaf.

6. Inoculate one explant piece per culture tube, with five replications of each explant (lamina or petiole) per treatment. Incubate culture tubes on slant racks at 18°–20°C under low-light conditions (60 μEM²/sec).

7. Approximately 60 days after inoculation, shoots which have reached a size of at least 15 mm can be transferred to a root induction medium. This is accomplished by aseptically cutting small numbers of shoots from the leaf disks using a scalpel and forceps. These small individual shoots or shoot clumps are then placed with the bottom of the shoot or stem touching the root induction medium. Shoots which have not reached this height can be transferred back onto a shoot induction medium.

8. Thirty to forty-five days after inoculation on the root induction medium, or when a well-established root system has been formed, remove the closure to the culture vessel to allow the shoots to begin the acclimatization process. Allow the cultures to remain like this for 7–14 days prior to transferring the rooted shoots to a soil mix medium. Once the plantlets are in a soil mix, place them in a greenhouse under intermittent mist for several days to allow for further acclimatization.

Scheduling

Event	Timing
Isolation of explants	Day 0
First signs of morphogenesis	ca. Day 14
Subculture of explants	Day 60–90

Recording Results
1. Record all details of setting up the experiments.
2. Make visual observations at 14-day intervals.
3. Determine fresh weight and shoot number of the explants between 60 and 90 days.

REFERENCES

1. Appelgren, M. 1976. Regeneration of *Begonia hiemalis in vitro*. Acta Hort. 64:31–38.

2. Hilding, A. S. and T. Welander. 1976. Effect of some factors on propagation of *Begonia × hiemalis in vitro*. Swed. J. Agric. Res. 6:191–199.

3. Khoder, M., P. Villemur, and R. Jonard. 1981. La multiplication vegetative de l'espece florale *Begonia elatior* (cultivar Rieger). a' partir de differents organes cultive's *in vitro*. C.R. Acad. Sci., Ser. III 293:403–408.

4. Linsmaier, E. M. and F. Skoog. 1965. Organic growth factor requirements for tobacco tissue culture. Physiol. Plant. 18:100–127.

5. Mikkelsen, E. P. and K. C. Sink. 1978. *In vitro* propagation of Rieger Elatior begonias. HortScience. 13:242–244.

6. Nitsch, J. P. and C. Nitsch. 1969. Haploid plants from pollen grains. Science 163:85–87.

7. Roest, S. 1977. Vegetative propagation *in vitro* and its significance for mutation breeding. Acta Hort. 78:349–359.

8. Simmonds, J. 1984. Induction, growth and direct rooting of adventitious shoots of *Begonia × hiemalis*. Plant Cell, Tissue and Organ Culture 3:283–289.

9. Takayama, S. and M. Misaivce. 1981. Mass propagation of *Begonia × hiemalis* plantlets by shake culture. Plant Cell Physiol. 22:461–467.

10. Welander, T. 1977. *In vitro* organogenesis in explants from different cultivars of *Begonia × hiemalis*. Physiol. Plant. 41:142–145.

11. Welander, T. 1979. Influence of medium composition on organ formation in explants of *Begonia × hiemalis in vitro*. Swed. J. Agric. Res. 9:163–168.

12. Zimmer, K. and H. Potthoff. 1978. Gervebeculture bei Begonien (1). Deutsche Gartenbau 21:867–869.

15

Sweet Potato Tissue Culture

The sweet potato, *Ipomoea batatas* (L.) LAM., is a member of the morning glory or beniveed family (Convolvulaceae), which contains about 50 genera and over 1200 species. The sweet potato is the only member of the genus *Ipomoea* whose root is edible.

Sweet potatoes are of prime importance to the food supply and economics of many countries. On a global basis, the sweet potato ranks sixth in annual production of major food crops (following wheat, rice, maize, potato, and barley); in Asia, it accounts for 62% of the total acreage devoted to root and tuber crop production (5, 11). Sweet potatoes are a good source of energy, supplying sugars and other carbohydrates; of calcium, iron, and other minerals; and of vitamins, particularly vitamins A and C (2). Approximately 80–85% of the sweet potatoes produced in the United States each year are used as human food (5).

Although the *in vitro* propagation of *I. batatas* has not been as successful or common as that of other vegetable crops, several reports concerning tissue culture of sweet potato have appeared in recent years. These contributions have dealt with callus induction in tissue explants followed by plantlet regeneration (1, 6, 7, 8) and with regeneration of plants by meristem culture (1).

SHOOT PROLIFERATION/ORGANOGENESIS

Litz and Conover (6) found that several sweet potato cultivars could be propagated *in vitro* by using lateral buds and shoot apices as the primary explant source. They reported that an average of 8.5 plantlets per culture was obtained from 'White Star' cultured on a modified MS medium supplemented with 1.0 mg/liter BA. An average of 5.1 plantlets per culture was obtained from the cultivar PI 315343 when 1.0 mg/liter kinetin and 1.0 mg/liter BA was added to a MS culture medium.

Rooted and unrooted plants were successfully rooted and transferred to greenhouse conditions. The authors observed no evidence of variability when the plants were transferred to field conditions.

Alconero et al. (1) reported that meristematic tips (0.4–0.8 mm long) of axillary shoots of sweet potato developed into complete plants in 20–50 days when cultured on a modified MS agar medium. They evaluated 10 cultivars of sweet potatoes and noted some differences in shoot development among them. Auxin source was quite significant; NAA produced only a few shoots and callus, whereas IAA resulted in higher shoot production. The best plantlet production was obtained with a combination of IAA and kinetin. The largest number of normal shoots formed when 2.0 mg/liter kinetin and 1.0 mg/liter IAA were included in the medium, but a larger number of plantlets was achieved with other combinations.

Templeton-Somers and Collins (12) evaluated the somaclonal variability of I. batatas 'Jewel' propagated in vitro. They noted that plants produced from slips and cuttings had higher yields than in vitro-propagated plants. The number of roots with flesh color mutations was also higher in plants propagated by slips or cuttings. The difference in yield was insignificant the following year, but the skin color mutations remained high the second year.

CALLUS INITIATION AND GROWTH

Callus has been successfully initiated from anthers, shoots, and roots of the sweet potato. The latter explant is generally chosen for callus initiation because of its abundance, ease of manipulation, and fleshy nature. Several media formulations, such as White's, Gamborg's, and Murashige and Skoog's, have been used for callus initiation, with MS or modifications thereof having been used most frequently.

Yamaguchi and Nakajima (16) used a modified White's medium to produce callus and adventitious roots and shoots. To obtain adventitious roots, 0.5 to 50.0 μM of either kinetin or zeatin were required, but these concentrations inhibited bud formation. Adventitious bud formation from tubers in some varieties was stimulated with low concentrations of 2,4-D combined with ABA. Antoni and Folquer (3), using both White's and Murashige and Skoog's medium supplemented with 2,4-D and coconut milk, obtained callus formation from three cultivars of sweet potato; however, no organogenesis was obtained from the callus. Sehgal (9) initiated callus from excised leaves of sweet potato using an MS medium supplemented with 2.3 μM 2,4-D plus 0.5 μM kinetin. This callus proliferated readily and when transferred to an MS medium supplemented with either IAA of NAA, root initiation occurred. When calli were transferred to an MS medium supplemented with 7.4 μM to 11.8 μM adenine sulfate or 0.5 to 2.3 μM kinetin, both roots and shoots were initiated.

EMBRYOGENESIS

Asexual embryos were initiated from callus derived from axillary bud shoot tips of six sweet potato plant lines (7, 8). Best results for embryogenesis were obtained on an MS medium supplemented with 2,4-D at 0.1 to 3.0 mg/liter. Transferring embryogenic callus to an auxin-free medium resulted in continued embryo development and eventual germination of the embryos.

Liu and Cantliffe (8) induced somatic embryos from callus derived from shoot tips when cultured on an MS medium supplemented with 0.5–3.0 mg/liter 2,4-D. Embryogenic callus could be proliferated with a medium supplemented with 2.0 mg/liter 2,4-D plus 0.25 mg/liter kinetin. When the embryogenic callus was transferred to a hormone-free medium, globular shaped embryos readily developed into torpedo-shaped embryos before germinating into whole plants.

Carswell and Locy (4) initiated roots, shoots, and callus from stem, leaf, and storage root tissue of the sweet potato cultivar Jewel. Best callus initiations were observed when the explants were cultured on a medium supplemented with 1.0 mg/liter NAA and 10.0 mg/liter BA. Shoot formation increased the longer the original explant remained in culture without being subcultured. The greatest proliferation of shoots occurred on a medium supplemented with 1.0 mg/liter NAA plus 0.1 mg/liter BA.

ANTHER CULTURES

Callus tissue has been derived from anthers of the sweet potato flower (10, 13–15). Tsai and Lin (14) used a Blaydes medium supplemented with IAA, 2,4-D, kinetin, and coconut milk to produce callus. Best callus induction occurred when 2,4-D was used as the auxin source, but best growth in subsequent passages occurred when 2,4-D and IAA were used together as the auxin sources. Shoot production did not occur in the treatments reported. Diploid plantlets were produced from anthers of sweet potato or from anther-derived callus; however, to date no true haploid plantlets have been obtained.

Materials Required
1. 10 sterile glass or plastic petri dishes (100 mm in diameter)
2. 500-ml beaker and 250-ml beaker
3. 3 pairs of forceps and 3 scalpels
4. Culture tubes (25 × 150 mm) and plastic racks for holding them
5. Waterproof marking pen
6. Bunsen or alcohol burner
7. 12 quart Mason jars
8. 1000 ml of 20% Clorox solution supplemented with a few drops of Tween-20
9. 1000 ml of sterile distilled water
10. 2 liters of basic culture medium containing the following components:

MS salts[1]
Sucrose	30.0 g/liter
Agar	8.0 g/liter
B-5 vitamins	1.0 ml/liter

11. 12 average-sized sweet potatoes from grocery store. This number will be required to produce enough shoots and explants.

Procedures
1. Place one sweet potato root in each Mason jar filled with water. Place the jars on a window sill or in an incubator at 25°C and allow the roots to sprout new shoots. Shoots of ample length should grow within 6–12 weeks.
2. Wipe down all surfaces of the transfer hood and turn on the hood for 15 min before doing any work under it.
3. Prepare basic culture medium as described in the previous section, divide it into two 1-liter portions, and add the following:

Treatment A		**Treatment B**	
NAA	0.1 mg	NAA	5.0 mg
BA	0.5 mg	BA	1.0 mg

Dispense 25 ml/tube of supplemented medium into tubes, with 40 replications per treatment, and sterilize.
4. Collect 10–20 sweet potato shoots and wash them under running tap water for 30 min. Cut shoots into sections about 10 cm in length, remove all leaves, place tissue in a beaker, and sterilize

[1] Prepare as described in Chapter 2 or use commercially prepackaged salts at the recommended rate.

with Clorox solution for 5 min. Following sterilization, decant the disinfectant and rinse the tissue three times with sterile distilled water.

5. Section the tissue so the explant to be cultured contains one node and is no more than 2.5 cm in length. Inoculate one explant per culture tube.

6. Incubate the cultures under low-light conditions (100 $\mu E/M^2/sec$) for 60 days and then evaluate the growth of the tissue at that stage.

Scheduling

Event	Timing
Isolation of the explant	Day 0
Appearance of bud break from lateral buds	Day 7–14
First subculture	Day 45

Recording Results

1. Record all details of setting up the experiments.
2. Make visual observations at 7-day intervals.
3. Calculate the number and size of the shoots at day 45.

REFERENCES

1. Alconero, R. A., G. Santiago, F. Morales, and F. Rodriguez. 1975. Meristem tip culture and virus indexing of sweet potatoes. Phytopathology 65:769–773.
2. Anon. 1980. Sweet Potato Quality. Southern Cooperative Series Bull. 249.
3. Antoni, H. J. and F. Folquer. 1975. In vitro tissue culture of sweet potatoes, *Ipomoea batatas* (L.) Lam., for the production of new cultivars. Rev. Agron. Noreste Argent. 12:177–178.
4. Carswell, G. K. and R. D. Locy. 1984. Root and shoot initiation by leaf, stem, and storage root explants of sweet potato. Plant Cell, Tissue and Organ Culture 3(3):229–236.
5. Jones, A. 1970. The sweet potato—today and tomorrow. pp. 3–6. *In* Tropical Root and Tuber Crops Tomorrow. Vol. I. Proc. of the Second Int. Symp. on Tropical Root and Tuber Crops.
6. Litz, R. E. and R. A. Conover. 1978. *In vitro* propagation of sweet potato. Hort-Science 13:659–660.
7. Liu, J. R. and D. J. Cantliffe. 1983. Somatic embryogenesis and plant regeneration in tissue culture of sweet potato (*Ipomoea batatas*). HortScience 18:618.
8. Liu, J. R. and D. J. Cantliffe. 1984. Improved efficiency of somatic embryogenesis and plant regeneration in tissue cultures of sweet potato (*Ipomoea batatas*). Hort-Science 19:588.
9. Sehgal, C. B. 1975. Hormonal control of differentiation in leaf cultures of *Ipomoea batatas*. Poir. Beitr. Biol. Pflanz. 51:47–52.
10. Sehgal, C. B. 1978. Regeneration of plants from anther culture of sweet potato (*Ipomoea batatas* Poir) Z. Pflanzenphysiol 88:349–352.

11. Steinbauer, C. E. and L. J. Kushman. 1971. Sweet Potato Culture and Diseases. Agric. Handb. *388*. U.S. Dept. Agriculture, Washington, DC.
12. Templeton-Somers, K. M. and W. Collins. 1984. Field performance and phenotypic stability of sweet potatoes propagated *in vitro*. HortScience 19:598.
13. Tsai, H. S. and C. I. Lin. 1973a. The growth of callus induced for in vitro culture of sweet potato anthers. J. Agric. Assoc. China 81:12–19.
14. Tsai, H. S. and C. I. Lin. 1973b. Effects of the composition of culture media and cultural conditions on growth of callus of sweet potato anther. J. Agric. Assoc. China 82:30–41.
15. Tsai, H. S. and M. T. Tseng. 1979. Embryoid formation and plantlet regeneration from anther callus of sweet potato. Bot. Bull. Acad. Sin. 20:117–122.
16. Yamaguchi, T. and T. Nakajima. 1972. Effect of abscisic acid on adventitious bud formation from cultured tissue of sweet potato. Crop Sci. Soc. Jpn. Proc. 41:531–532.

16

Establishment of Root Cultures
In Vitro

One of the pioneering events in the development of tissue culture techniques was the successful establishment of actively growing clones of tomato roots reported by P. R. White in 1934. In this classic research, White cultured tomato roots through 160 successive passages, each 7 days in duration. During this period, the tomato roots, when cultured on an optimal medium, grew an average of 5 mm/day. The importance of several B vitamins and of auxin for cell growth was recognized as a result of White's work with tomato roots.

The nutrient medium described by White (4) consisted of inorganic salts, sucrose, and yeast extract. Street (3) later expounded the detailed nutritional requirements of tomato root cultures. The primary medium used for root cultures has been that of White, however, iron deficiencies were noted as root growth slowed when the pH of the medium was above 5.2. The use of Fe–EDTA results in the increased availability of iron over a wider pH range thus eliminating iron deficiencies. Sucrose is generally the carbon source of choice for most root cultures. Sucrose concentrations of 15–20 g/liter are adequate with higher or lower concentrations resulting in shortened roots with few if any laterals (1, 3). The gentle agitation of the root culture may lead to a doubling in root elongation and an increase in the number of lateral roots produced (2). A list of early research with root cultures was prepared by Butcher and Street (1).

Modern methods for culturing excised root tips are basically the same as those described by White. Most methods involve the germination of seed aseptically, removal of the radical upon emergence, and transfer to liquid medium. Once root growth begins and becomes active, the terminal 10 mm of the root and root laterals can be excised and placed on fresh medium.

Materials Required
1. 6 Erlenmeyer flasks (250 ml)
2. Sterile plastic petri dishes (100 × 15 mm)

3. 3 scalpels and 2 pairs of forceps
4. Waterproof marking pen and stick-on labels
5. Ethanol or bunsen burner
6. 400 ml of 95% ethanol
7. 400 ml of 1% aqueous solution of Tween-20
8. 400 ml of 20% Clorox solution
9. 2000 ml of sterile distilled water
10. Culture medium consisting of the following components:

MS medium	Amount
MS salts[1]	
Sucrose	15.0 g/liter
Thiamin HCl	0.5 mg/liter
Myo-inositol	100.0 mg/liter

White's medium	Amount
White's basal salts[1]	
Sucrose	15.0 g/liter
Thiamin HCl	0.5 mg/liter
Myo-inositol	100.0 mg/liter

11. About 50 dry, undamaged pea seed (*Pisum sativum*)
12. About 50 dry, undamaged tomato seed

Procedures

1. Place the pea seed in a 250-ml Erlenmeyer flask. Pour 200 ml 95% ethanol over the seed and shake for 10 sec, then pour off the ethanol (see Fig. 16.1). Pour 200 ml of the Tween-20 solution over the seed and shake the flask gently for 1 min. Examine the seed individually and discard those that float or appear to have a translucent tissue beneath the seed coat. Sterilize the remaining seed by pouring 200 ml of Clorox solution over the seed and disinfect for 20 min under aseptic conditions. Following sterilization, wash the seed three times in sterile distilled water, using about 200 ml per rinse.

2. Pour 20 ml of sterile distilled water into each of 10 petri dishes and place 2 seeds in each dish. Incubate the plates in the dark at 25°C for 48 hr.

3. Prepare basic culture medium as described in previous section

[1] Prepare as described in Chapter 2 or use commercially prepackaged salts at the recommended rate.

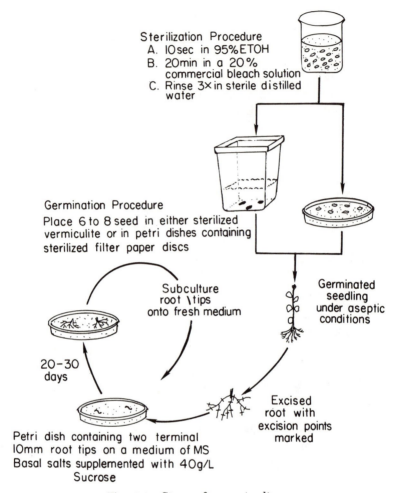

Sterilization Procedure
A. 10 sec in 95% ETOH
B. 20 min in a 20% commercial bleach solution
C. Rinse 3× in sterile distilled water

Germination Procedure

Place 6 to 8 seed in either sterilized vermiculite or in petri dishes containing sterilized filter paper discs

Subculture root \tips onto fresh medium

Germinated seedling under aseptic conditions

20–30 days

Excised root with excision points marked

Petri dish containing two terminal 10mm root tips on a medium of MS Basal salts supplemented with 40g/L Sucrose

Fig. 16.1. Stages of pea root culture.

and sterilize it. Under aseptic conditions, dispense 20 ml of medium into each of 10 sterile petri dishes. Repeat procedure for both media.

4. When the primary root of each pea seedling (step 2) reaches 20–25 mm, excise the terminal 10 mm and transfer the tip to a plate containing basic medium. Inoculate two tips per plate. Repeat this procedure until all culture plates have been inoculated.

5. Incubate all cultures at 25°C in the dark.

6. Repeat steps 1 and 2 using tomato seed.

7. Supplement the basic culture medium with either 0.0 mg/liter

2,4-D (treatment A), 0.01 mg/liter 2,4-D (treatment B), or 1.0 mg/liter 2,4-D (treatment C). Under aseptic conditions, dispense 20 ml/plate of basic medium and supplemented media (A and B) into sterile petri dishes. Prepare 5 replicates per treatment.

8. Remove primary root tips from tomato seedlings as described in step 4 and inoculate onto culture plates prepared in step 7. Incubate cultures at 25°C in the dark.

9. Subculture half of the cultures at 7-day intervals. Do not subculture the remaining cultures.

Scheduling

Event	Timing
Sterilization of pea and tomato seed and incubation under aseptic conditions	Day 0
Removal of seedling's primary root and reculture on media	Day 2–4
Daily measurement of root growth	Day 3–30
First appearance of lateral roots	Day 5–10
Subculture of roots	Day 7, 14, 21, and 28
Termination of experiment	Day 30

Recording Results

1. Record all procedures required in setting up the experiment.
2. Aseptically measure root length once every 24 hr either by removing the root and placing it on a sterile ruler or by placing the petri plate on a piece of graph paper with divisions in millimeters.
3. Plot your results graphically.
4. Determine wet and dry weights of all root cultures at the end of the 30-day culture period.
5. Visually rate any callus formation.

REFERENCES

1. Butcher, D. N. and H. E. Street. 1964. Excised root culture. Botanical Rev. 30(4): 513–586.
2. Said, A. G. E. and T. Murashige. 1979. Continuous cultures of tomato and citron roots *in vitro*. In Vitro 15:593–602.
3. Street, H. E. 1957. Excised root culture. Biol. Rev. 32:117–155.
4. White, P. R. 1963. The Cultivation of Plant and Animal Cell. 2nd ed. Ronald Press, New York.

17

Effects of Carbohydrates, Vitamins, and Gelling Agents on Callus Growth and Plantlet Regeneration

As noted in earlier chapters, the composition of the medium, as well as the physical characteristics of the environment (e.g., temperature, light, humidity), greatly influences the growth and morphogenesis of plant tissues during *in vitro* culture. The effects of growth regulators have been evaluated in several of the previous chapters. In this chapter, the effects of carbohydrates, vitamins, and gelling agents on callus growth and plant regeneration are briefly reviewed and demonstrated.

CARBOHYDRATES

Carbohydrate requirements for callus maintenance were initially investigated by Gautheret (9) and White (30). The superiority of sucrose as the carbohydrate source was first reported by White (33) and Dormer and Street (5) and has been confirmed by numerous workers (4, 6, 7, 8, 10, 12, 14, 17, 24, 26, 28). Other carbohydrate sources also have been shown to adequately support tissue growth in culture (6, 7, 15, 16, 17, 28). It has been suggested that the carbohydrates used to support growth may be metabolized through a common intermediate, which may be produced at different rates depending upon the carbohydrate source (28). Thus, the rate of uptake and conversion of the carbohydrate source to the intermediate could regulate growth.

Glucose has been substituted for sucrose without a loss of cell growth in haplopappus (*Haplopappus gracilis*) (7), and a wide variety of carbohydrates have been shown to support embryo formation in carrots (28). Fructose, glucose, and trehalose supported growth as well as sucrose in cell suspension cultures of sugarcane (17) and Paul's scarlet rose (16). In annual ryegrass, glucose and fructose supported growth equal to that obtained with sucrose during the first generation; however, during the second generation, callus growth was greater with sucrose than with fructose or glucose.

Carbohydrate concentration has also been shown to influence callus maintenance. Sucrose concentrations between 20 and 60 g/liter have been used to support callus growth (7, 8, 9, 12, 16, 17, 23, 24, 28, 30, 31). Callus cultures of tobacco have been shown to be as dependent on carbohydrate concentration as on the hormonal regime (11). The optimal sucrose concentrations for in vitro growth of carrot tissues has been reported to be 3% (8).

The method of sterilizing (autoclaved vs. filter sterilization) specific carbohydrates also affects cell growth in culture. In particular, the use of autoclaved fructose results in decreased cell growth in several plant species (1, 6, 16, 18). Stehsel and Caplin (23) compared the effects of autoclaved vs. filter-sterilized fructose on carrot root tissue cultures. They found that autoclaved fructose caused both an inhibition of callus growth and a noticeable discoloration of callus. The degree of inhibition caused by autoclaved fructose increased with increasing fructose concentrations from 1 to 2% and was greatest when fructose was autoclaved in the presence of coconut milk. These authors reported similar but less pronounced effects with autoclaved glucose; however, both fructose and glucose, when filter-sterilized, supported adequate growth (23). Autoclaved fructose has also been reported to inhibit callus growth in Paul's scarlet rose (Rosa sp.); the most significant inhibition occurred when fructose was autoclaved in the presence of micronutrients, specifically magnesium sulfate (16).

Materials Required
1. 5 sterile glass or plastic petri plates (100 mm in diameter)
2. 250-ml beaker
3. 2 scalpels with new blades, 3 pairs of forceps, and 3 spatulas
4. Waterproof pen and stick-on labels
5. Bunsen or ethanol burner
6. Culture tubes (20 × 150 mm) and plastic racks to hold them
7. 150 ml of 95% ethanol
8. 250 ml of 20% Clorox solution supplemented with a few drops of Tween-20
9. 300 ml sterile distilled water
10. Stock callus cultures from one or two species of plants (lily or 'Penncross' creeping bentgrass work well)
11. 10 medium-sized healthy African violet leaves

Procedures
1. Prepare 4 liters of the following medium.

MS basal salts[1]
Agar	8.0 g/liter
Thiamin HCl	1.0 mg/liter
Myo-inositol	100.0 mg/liter
NAA	0.1 mg/liter
BA	1.0 mg/liter
pH	5.7

This media will be used for the regeneration studies.
2. Prepare 4 liters of the following medium.

MS basal salts[1]
Agar	8.0 g/liter
Thiamin HCl	1.0 mg/liter
Myo-inositol	100.0 mg/liter
2,4-D	2.5 mg/liter
BA	0.1 mg/liter
pH	5.7

This medium will be used for the callus maintenance studies.
3. Divide each of the media into 250-ml aliquots. To each of the aliquots add one of the following treatments:

0.0 g/liter sucrose	0.0 g/liter glucose
20.0 g/liter sucrose	20.0 g/liter glucose
40.0 g/liter sucrose	40.0 g/liter glucose
60.0 g/liter sucrose	60.0 g/liter glucose
0.0 g/liter maltose	0.0 g/liter galactose
20.0 g/liter maltose	20.0 g/liter galactose
40.0 g/liter maltose	40.0 g/liter galactose
60.0 g/liter maltose	60.0 g/liter galactose

Add 10 ml of the above media treatments per culture tube. This will give you 25 replications per treatment or a total of 400 tubes for each of the media that you prepared. Label all tubes with treatment, date, medium, and tissue to be inoculated into the culture tube. Prepare African Violet leaf disks using the procedures described below for the regeneration medium. Prepare callus cultures as described below and inoculate onto the callus maintenance medium.

---

[1] Prepare as described in Chapter 2 or if using commercially prepackaged salts use as directed on the package.

Callus Cultures
1. Under aseptic conditions, remove a portion of callus tissue from stock callus culture of creeping bentgrass or lily. Cut into small pieces, about 2 mm in diameter.
2. Weigh each callus piece and then inoculate into culture tubes (one piece per tube) containing media prepared in the previous section.
3. Incubate the culture tubes at 25°C in the dark. Periodically monitor the growth of the callus in each treatment.

African Violet Leaf Cultures
1. Place enough leaves to complete the project in a beaker and wash gently under running tap water. Sterilize the leaves for 6–8 min in 20% Clorox solution, then rinse three times with sterile distilled water. Make certain that all leaves are thoroughly covered with each rinse.
2. Remove all outer edges of the leaves and section the remaining portion into 1-cm squares. Inoculate one leaf section, adaxial side down, per culture tube.
3. Incubate the culture tubes at 25°C in low light. Periodically monitor the regeneration of plantlets from leaf pieces.

Scheduling

Event	**Timing**
Isolation and inoculation of explants	Day 0
First appearance of plantlet formation	Day 14–21
First noticeable effects of carbohydrate treatments	Day 21–28
Termination of experiment	Day 60

Recording Results
1. Record all details of setting up the experiment.
2. Make visual observations at 14-day intervals.
3. Determine fresh weight gain of callus cultures and shoot number of leaf and callus cultures after 60 days.

VITAMINS

Several vitamins—including thiamin, pyridoxine, nicotinic acid, folic acid, ascorbic acid, and *myo*-inositol—have been shown to be essential for proper growth, metabolism, and division of plant cells in culture (4, 14, 19). The B-complex vitamins are particularly important, and thiamin (vitamin B_1) is especially critical in most species. Thiamin functions as a coenzyme in the decarboxylation of keto acids and is important in cellular metabolism in the whole plant. The omission of thiamin from the medium has been shown to reduce callus growth in most species (4, 14). Yeast extracts used in earlier media have been replaced by thiamin and pyridoxine HCl in some species (20, 21, 22, 31, 32). The minimum concentration of thiamin required for satisfactory callus growth of tobacco cultivars was 400 μg/liter (14).

Nicotinic acid influences carbohydrate metabolism, meristematic activity, and cell dimension in *in vivo* cultures. However, in tobacco callus cultures, the addition of nicotinic acid showed no beneficial effects and decreased yields at the higher concentrations tested (14).

In whole plants, pyridoxine (vitamin B_6) is involved in the enzymatic reactions dealing with amino acid metabolism and is a coenzyme in amino acid decarboxylation reactions. In tissue culture, lack of pyridoxine has been shown to decrease meristematic activity of roots (2, 3, 4). Pyridoxine has been reported to have little effect on yield or vigor of tobacco callus (14).

In whole plants, folic acid functions as an intermediate carrier of one-carbon units. Linsmaier and Skoog (14) found that folic acid increased callus yields and improved color in tobacco cultures in the light but inhibited callus growth in the dark.

Ascorbic acid functions in whole plants as a catalyst in photosynthetic phosphorylation and may serve in electron transfer from NADPH to oxygen by undergoing cyclic oxidation-reduction. Ascorbic acid (vitamin C) had no effect on carrot callus yields; however, its omission resulted in an increase in cell number (27). Ascorbic acid added to the medium was found to stop melanin formation, which has been shown to be detrimental to callus cultures (4). Callus yields increased and color improved with the addition of ascorbic acid to tobacco culture media (14).

In whole plants, myo-inositol is involved in the synthesis of cell wall polysaccharides, and in the uptake and utilization of ions. It is also present in the lipid fraction (inositol-phosphosphigolipid) of cell membranes (13). Deletion of myo-inositol had no significant effect on callus yields in carrots and hapolpappus but resulted in decreased fresh weight gains in tobacco cultures at all levels of thiamin (14). Myo-inositol has also been reported to promote bud formation in elm (*Elmus* sp.) cultures. Presently, only a limited number of reports are available on the effects of nicotinic acid, folic acid, and ascorbic acid on plant tissue cultures.

Materials Required
1. 5 sterile glass or plastic petri plates (100 mm in diameter)
2. 250-ml beaker
3. 2 scalpels with blades and 3 spatulas

4. Waterproof pen and stick-on labels
5. Bunsen or ethanol burner
6. Culture tubes (25 × 150 mm) and slant racks for holding them
7. Stock callus cultures of 4 creeping bentgrass cultivars

Procedures
1. Prepare 3 liters of the following medium.

MS basal salts[1]

Sucrose	30.0 g/liter
Agar	8.0 g/liter
2,4-D	5.0 mg/liter
BA	1.0 mg/liter
Myo-inositol	100.0 mg/liter

2. Divide the media into 12 equal aliquots of 250 ml each. To each of the aliquots add one of the following treatments:

0.0 mg/liter thiamin	0.0 mg/liter nicotinic acid	0.0 mg/liter pyridoxine
0.1 mg/liter thiamin	0.1 mg/liter nicotinic acid	0.1 mg/liter pyridoxine
1.0 mg/liter thiamin	1.0 mg/liter nicotinic acid	1.0 mg/liter pyridoxine
10.0 mg/liter thiamin	10.0 mg/liter nicotinic acid	10.0 mg/liter pyridoxine

3. Adjust the pH of each treatment then dispense 10 ml of media into 20 × 150 mm culture tubes, cap, label each treatment, then sterilize at 121°C, 1 kg/cm^3 for 17 min. There should be 25 culture tubes per treatment.
4. Best results are obtained in these studies if the callus to be inoculated for the study is "pulsed" by placing the callus on the medium listed below 2–4 weeks prior to the initiation of the study.

MS basal salts[1]

Sucrose	30.0 g/liter
Agar	8.0 g/liter
2,4-D	5.0 mg/liter
BA	1.0 mg/liter

The purpose of this is to allow the cells to utilize any vitamins "stored" within the tissue thus giving a more accurate indication of the vitamin effects.

[1] Prepare as described in Chapter 2 or if using commercially prepackaged salts use as directed on the package.

Callus Cultures
1. Under aseptic conditions, remove a portion of callus tissue from each of the bentgrass stock cultures (some from each cultivar). Cut into small pieces, about 3 mm in diameter. You will need 25 pieces/treatment.
2. Weigh each callus piece and then inoculate into culture tubes containing media prepared in the previous section, one piece per tube.
3. Incubate cultures at 25°C in the dark. Periodically monitor the growth of the callus in each treatment.

Scheduling

Event	Timing
Inoculation of explants	Day 0
First appearance of treatment effects	Day 21–28
Termination of experiment	Day 60

Recording Results
1. Record all details of setting up the experiments.
2. Make visual observations at 14-day intervals.
3. Determine fresh weight of calli after 60 days.

GELLING AGENTS

Because it is usually desirable for tissue explants to be in contact with but not submerged in the culture medium, solid or semisolid media often are used. Tissues cultured on such media maintain good contact with the media but also have good aeration. Agar, gelatin, starch-derivative gels, and silica gels have all been used to solidify culture media. Agar, however, is the most commonly used gelling agent in plant tissue culture work.

The agar used to solidify culture media is available in various grades, the most highly purified grades often being very expensive. Commonly, the agar used for defined culture media is purified by an ion-exchange process. Several commercial agar preparations are available; Difco Noble bacteriological agar is the purest. The other grades are usually acceptable for routine work where the presence of small amounts of interfering substances is not considered important. Agar can be purified further in the laboratory, but this usually is not necessary unless critical micronutrient work is planned (25). Agar is widely available as a dry powder, obtained by extracting and drying mucilaginous substances from the seaweed *Gelidium*. When dissolved at concentrations ranging from 0.6 to 1.0% (i.e., 0.6 to 1.0 g/100 ml) in water, it forms a semisolid gel suitable for *in vitro* culture. To dissolve agar, it is necessary to heat the culture medium to 98°C (208°F). Solidification of the agar medium occurs at approximately 40°C (104°F), but the solidification temperature varies with the grade and purity of the agar. As long as an agar solution is kept hot, it can be poured, pipetted, or otherwise transferred to culture vessels. Once solidified, it must be reheated to 98°C before it liquifies again (25, 29).

Materials Required
1. Sterile plastic petri plates (15 × 100 mm)
2. 250-ml beaker
3. 2 scalpels with blades and 3 spatulas
4. Waterproof pen and stick-on labels
5. Bunsen or ethanol burner
6. Culture tubes (25 × 100 mm) and slant racks for holding them
7. 150 ml of 95% ethanol
8. 400 ml of 20% Clorox solution supplemented with a few drops of Tween-20
9. Difco Bacto Agar
10. Difco Noble Agar
11. K. C. Biological TC Agar
12. Gibco Phytar Agar
13. Sigma Chemical Plant TC Agar
14. Sigma Phytagel (Use at 2g/liter concentration)
15. 1 small to medium-sized lily bulb
16. Healthy African violet plants grown under greenhouse conditions
17. Stock callus cultures of 'Penncross' creeping bentgrass
18. Actively growing carrot (Daucus) cell suspension cultures

Preparation of Media
1. Prepare 6 liters of the basal medium (BM) listed below:

MS salts[1]	
Sucrose	30.0 g/liter
Thiamin HCl	1.0 mg/liter
Myo-inositol	100.0 mg/liter

2. Divide the basic medium into 1.5-liter portions and supplement according to the following protocol:

---

[1] Prepare as described in Chapter 2 or if using commercially prepackaged salts use as directed on the package.

	Supplement	Conc. (mg/liter)
Lily propagation medium	NAA	0.1
	BA	0.1
African violet propagation medium	NAA	0.1
	BA	0.1
Bentgrass callus medium	2,4-D	5.0
	BA	0.1
Carrot cell medium	NAA	1.0
	Kinetin	0.1

3. Divide each of the four hormone-supplemented media into six 250-ml aliquots. Next, add 2.0 g of each type of agar to one 250-ml aliquot of each media except the carrot medium where 4.0 g/liter is required. Adjust the pH of all media to 5.7 ± 0.1 with either NaOH or HCl. Once the agar is completely dissolved, pour the media into culture tubes (25 ml/tube). There should be 10 tubes/agar treatment for each plant species. Label all tubes and then sterilize. Handling of carrot media is described later.

Lily Propagation
1. Remove outer damaged scales, basal plate, and innermost scales from the bulb.
2. Wash scale sections in warm, soapy water then rinse under running water for 15 min. Sterilize for 2 min in 95% ethanol and then for 15 min in 20% Clorox solution. Rinse the scales three times in sterile distilled water.
3. Section bulb scales into 1- to 2-mm sections. Inoculate one section per culture tube. Incubate at 25° ± 1°C under low light.

African Violet Propagation
1. Remove small to medium actively growing leaves from African plants.
2. Wash leaves under running water; disinfect for 8 min in 20% Clorox solution. Rinse three times in sterile distilled water.
3. Remove the outer edges of the leaf and section the remaining portion into 1-cm² sections. Inoculate one section per culture tube. Incubate at 25° ± 1°C under low light.

Bentgrass Callus Maintenance
1. Obtain several small flasks that contain actively growing callus of *Agrostis palustrus* Huds. 'Penncross'. Cultures should have been last subcultured within 60 days.
2. Remove calli from flasks and place in sterile plastic petri dishes. Section all calli into small (2–4 mm in diameter) sections. Inocu-

late one section per culture tube. After determining fresh weight of the inocula, incubate at 25° ± 1°C under low light.

Carrot Cell Maintenance

1. Obtain several small flasks of actively growing cell suspensions of *Daucus*. Gently spin cell suspension down, remove the old medium, and wash the cells with fresh medium. Repeat the wash procedure three times.
2. Following the final wash, resuspend the cells in fresh carrot culture medium minus agar. Allow the carrot medium containing agar to cool to 40°C then mix equal volumes of the cell suspension with the carrot medium containing agar in a petri plate. Swirl to mix the suspension; then allow the media to solidify. This will give a final agar concentration of 8.0 g/liter.
3. Seal plates with Parafilm and label them. Incubate at 25° ± 1°C under low light.

Scheduling

Event	Timing
Inoculation of explants	Day 0
Termination of experiment	Day 60

Recording Results

1. Record all details of setting up the experiment.
2. Record visual observations at 14-day intervals or as required.
3. Determine fresh weight and shoot number and record appearance of tissues after 60 days.

REFERENCES

1. Ball, E. 1955. Studies of the nutrition of the callus cultures of *Sequoia sempervirens*. Ann. Biol. 31:80.
2. Boll, W. G. 1954. Investigations into the function of pyridoxine as a growth factor for excised tomato roots. Plant Physiol. 29:325.
3. Boll, W. G. 1954. The role of vitamin B_6 and the biosynthesis of choline in excised tomato roots. Arch. Biochem. Biophys. 53:20.
4. Butenko, R. G. 1964. Plant tissue culture and plant morphogenesis. Translated from Russian. Israel Program for Scientific Translation, Jerusalem, 1968.
5. Dormer, K. J. and H. E. Street. 1949. The carbohydrate nutrition of tomato roots. Ann. Bot. (London) 13:199.
6. Dougall, D. K. 1972. Cultivation of plant cells. *In* Growth, Nutrition and Metabolism of Cells in Culture. G. H. Rothblat and V. I. Cristafaol (editors). Academic Press, New York.

7. Eriksson, T. 1965. Studies on the growth requirements and growth measurements of cell cultures of *Haplapappus gracilis*. Physiol. Plant.

8. Gautheret, R. J. 1941. Action du saccharose sur la croissance de tissus de carotte. C. R. Soc. Biol. Paris 135:875.

9. Gautheret, R. J. 1955. The nutrition of plant tissue cultures. Annu. Rev. Plant Physiol. 6:433.

10. Goris, A. 1954. Transformations glucidiques intratissularies. Ann. Biol. 30:297–318.

11. Halgeson, J. P., C. D. Upper, and G. T. Haberback. 1972. Medium and tissue sugar concentrations during cytokinin-controlled growth of tobacco callus tissues. pp. 484–492. *In* Plant Growth Substances. 1970 Proc. 7th Int. Conf. on Plant Growth Substances. D. J. Carr (editor). Springer-Verlag, New York.

12. Hilderbrandt, A. C. and A. J. Riker. 1953. Influence of concentrations of sugars and polysaccharides on callus tissue growth *in vitro*. Am. J. Bot. 40:66.

13. Kaul, K. and P. S. Sabharwal. 1975. Morphogenetic studies on haworthia: Effects of inositol on growth and differentiation. Am. J. Bot. 62:655–659.

14. Linsmaier, E. M. and F. Skoog. 1965. Organic growth factor requirements of tobacco tissue cultures. Physiol. Plant. 18:100–127.

15. Mathes, M. C., M. Mariafrancce, and J. W. Marvin. 1973. Use of various carbon sources by isolated maple callus cultures. Plant and Cell Physiol. 14:797–801.

16. Nash, D. T. and W. G. Boll. 1975. Carbohydrate nutrition of Paul's scarlet rose cell suspension. Can. J. Bot. 53:179–185.

17. Nickell, L. G. and A. Maretzki. 1970. The utilization of sugars and starch as carbon sources by sugarcane cell suspension cultures. Plant and Cell Physiol. 11:183–185.

18. Redei, G. P. 1974. 'Fructose effect' in higher plants. Ann. Bot. 38:287–297.

19. Robbins, W. J. 1951. Vitamin and amino acid requirements for growth of higher plants. pp. 463–476. *In* Plant Growth Substances. Univ. of Wisconsin Press, Madison.

20. Robbins, W. J. and M. B. Schmidt. 1939. Further experiments on excised tomato roots. Bot. Gaz. 99:671–678.

21. Robbins, W. J. and M. B. Schmidt. 1939. Growth of excised tomato roots in synthetic solution. Bull. Torrey Bot. Club. 66:193–200.

22. Robbins, W. J. and M. B. Schmidt. 1939. Vitamin B_6, a growth substance for excised tomato roots. Proc. Nat. Acad. Sci. 25:1–3.

23. Stehsel, M. L. and S. M. Caplin. 1969. Sugars—autoclaving versus sterile filtration on the growth of carrot root tissues in culture. Life Sci. 8(2):1255–1259.

24. Street, H. E. 1966. Nutrition and metabolism of plant tissue and organ cultures. pp. 533–629. *In* The Biology of Cells and Tissues in Culture. Volume 3. E. N. Willmer (editor). Academic Press, New York.

25. Street, H. E. 1973. Plant tissue and cell culture. Blackwell Scientific Publications, Oxford.

26. Thomas, D. R., J. S. Craigie, and H. E. Street. 1963. Carbohydrate nutrition of excised tomato root. p. 2643. *In* Plant Tissue and Organ Culture. Proc. Symp. Int. Soc. Plant Morphogenesis, Delhi, India.

27. Torrey, J. G. and J. Reinert. 1961. Suspension cultures of higher plant cells in synthetic media. Plant Physiol. 36:483–491.

28. Verma, D. C. and D. K. Dougall. 1977. Influence of carbohydrates on quantitative aspects of growth and embryo formation in wild carrot suspension cultures. Plant Physiol. 59:81–85.

29. Wetherell, D. F. 1982. Introduction to *in vitro* propagation. Avergy's Plant Tissue Culture Series. Avery Publishing Group, Wayne, NJ.

30. White, P. R. 1934. Potentially unlimited growth of excised tomato root tips in liquid medium. Plant Physiol. 9:585–600.
31. White, P. R. 1937. Separation from yeast of the materials essential for growth of excised tomato roots. Plant Physiol. 12:777–791.
32. White, P. R. 1937. Vitamin B_1 in the nutrition of excised tomato roots. Plant Physiol. 12:803.
33. White, P. R. 1940. Does "C.P. Grade" sucrose contain impurities significant for the nutrition of excised tomato roots? Plant Physiol. 15:349–354.

PART III

Cell Culture

18

Overview of Cell Suspension Culture

A suspension culture consists of cells and cell aggregates dispersed and growing in a moving liquid medium. During incubation, the amount of cell material increases for a limited period of time until the culture reaches a point of maximum yield of cell material. If the culture is diluted at this point (subcultured), a similar pattern of growth and yield will occur again.

Suspension cultures are usually initiated by placing pieces of friable callus in an agitated liquid medium. Suspension cultures also can be started from sterile seedlings or embryos. Soft callus can be broken up in a hand-operated glass homogenizer; the homogenate (living, dead cells, and debris) is then transferred to the liquid culture medium. The first passage inoculum will contain residual pieces of the initial inoculum as well as more or less finely dispersed aggregates and free cells. It is recommended that suspensions be transferred using a pipette or syringe with an orifice small enough to exclude any large aggregates or residual pieces of inoculum.

The ideal cell suspension culture is characterized by morphological and biochemical homogeneity. However, most long-established cultures exhibit genetic diversity in their cell populations, usually as a result of differences in chromosome number or chromosome morphology. Such heterogeneity cannot be eliminated from such a mixed cell population.

CULTURE SYSTEMS

The Steward apparatus (7) was designed for rapid proliferation of small explants (3 mg) of carrot roots. The current vessel (tumble tube) is approximately 12.5 cm long and 3.5 cm in diameter. The tubes are run at 1–2 rpm. As the medium moves from end to end, the explant is alternately exposed to the medium and air (at 1 rpm the explant spends two-thirds of the time exposed to air).

In a similar device developed by Steward and Shantz (8), the culture tubes were replaced by *nipple flasks*. The slow movement of this device distributes the culture medium as a thin film over a considerable area within the culture vessel, thus allowing for a high rate of gaseous exchange.

Platform-orbital shakers consist of conical Erlenmeyer flasks placed on a platform shaker, which moves in a circular motion. Platform shakers have been widely used for the initiation and propagation of liquid cultures. By fitting clips on the platform, flasks from 50–2000 ml can be attached and shaken. Orbital shakers should have a variable speed control of 30–150 rpm. Speeds above 200 rpm are generally unsuitable for plant cell cultures. The stroke of the shaker should be between 3/4 and 1 1/2 inches orbital motion for optimal agitation. The optimum shaking speed (as judged by growth rate, total cell yield, and best cell separation) depends upon the particular culture, culture medium, type of culture vessel, and the volume of culture relative to vessel size and shape. Maintenance of stock cell suspensions generally requires using 20–25 ml of medium per 100-ml flask or 70 ml of culture medium per 250-ml flask.

Spinning cultures were first described in 1964 by Lamport (5). In this system, 10-liter culture vessels, holding 4.5 liters of medium, are placed on a culture platform at a 45° angle and rotated at 80–120 rpm. The yield with this type of system is comparable to that obtained with smaller (100–250 ml) flasks on a platform shaker.

Stirred and/or continuous culture systems are used for large batch cultures (volumes of 5–10 liters). In these systems, cells are maintained and distributed throughout the culture, and gas exchange is accomplished by either forced aeration or aeration with internal magnetic stirring. Stirred cultures may be easily hooked up to instruments for monitoring cell growth and media changes and connected to media reservoirs and gas supplies for altering cultural conditions. Systems such as this can be developed into continuous culture systems in which the temperature of the culture medium can be accurately controlled with an internal coil. The advantages of such systems include (1) ease of maintaining sterility over a long period of time, (2) freedom from mechanical failures during long periods of operation, (3) a degree of automation, (4) versatility in regards to growth conditions such as temperature, aeration, stirring or aeration speed, illumination, and nutrient and growth regulator levels, and (5) space for monitoring equipment.

CULTURE MEDIA

The general facilities and techniques for media preparation have been discussed in Chapters 1 and 2. Many media have been utilized for cell suspension cultures; however, most of the media that have been used to support callus growth generally are not for cell suspension cultures. Torrey and Reinert (10) showed that the vitamin requirements for rapidly growing cell suspensions differed from those for tissue (callus) cultures of the same clone. It, therefore, may be necessary to use a different culture medium for cell suspension cultures than for callus cultures.

Many media have very little buffering capacity, and the pH can change considerably once cells are introduced. Thus, monitoring and adjustment of pH may be necessary in a suspension culture.

Callus cultures may have to be subcultured frequently to maintain them in an actively growing state. Some callus cultures, however, can remain viable for weeks on a culture medium even though they are growing slowly. This is probably due to slow diffusion of nutrients from remote parts of the mass of solidified medium. Because cell suspension cultures may reach a very high cell density per unit volume culture (40–60% of the culture volume represented by cells is not uncommon), complete exhaustion of essential nutrients may occur by the time the culture enters the stationary phase.

GROWTH AND SUBCULTURING OF SUSPENSION CULTURES

When the cell number in suspension cultures is plotted against time of incubation a curve of growth is obtained, as shown in Fig. 18.1. Cell suspensions are clonally maintained by subculturing them in the early stationary phase. Beyond this point, the viability of the cell suspension may decrease. If a test of viability shows that viability is well-maintained over several days, this gives one more latitude in the timing of subcultures.

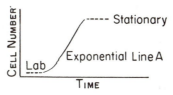

Fig. 18.1.   Growth curve.

The period from culture initiation to the stationary phase is determined primarily by: (1) initial cell density, (2) duration of lag phase, and (3) growth rate of the cell line. Usually cultures are established with a relatively high initial inoculum rate (0.5–2.5 × 10^5 cells). The cell number should rise during incubation to a range of 1–4 × 10^6 cells/ml before the culture reaches the stationary phase. This increase in the total number of cells is equivalent to every cell in the initial inoculum undergoing four to six cell divisions. For many clones, this will occur within 18–25 days; thus, the normal incubation time of stock cultures is 21–28 days between subcultures. When the cells are in a very active state of division, the passage length or periods between subcultures may be reduced to 6–9 days.

The use of low initial cell densities will prolong the lag and exponential phases of growth. This is usually recommended when one is trying to determine optimal medium for maximum growth. However, for each clone/culture medium combination there exists a critical initial density below which the culture will not grow. Using a standard synthetic culture medium, the critical initial density is 9–15 × 10^3 cells/ml. At this density, the cells will generally undergo eight doublings in cell number, reaching a maximum cell density before entering the stationary phase.

Conditioning of Medium

The critical or minimum initial cell density from which a new cell suspension can be reproducibly grown is a function of (1) the culture clone, (2) the duration and incubation of the stock culture being used to start the suspension, and (3) composition of the culture medium.

When a low initial inoculum density is used, cells can be induced to divide by means of a nurse culture or "feeder" cells. This indicates that the survival of single-cell clones or low-density cultures depends on substances that are not found in the normal culture medium but are released into the medium from the biosynthetic activity of the nurse cells. The metabolites released from the nurse cells alter the medium enough to stimulate the growth of low-density or single-cell cultures.

Stuart and Street (9) demonstrated that by using a "conditioned" medium, the initial cell density could be lowered by a factor of 10 to 1.0–1.5 × 10^3 cells/ml. A conditioned medium is prepared in an apparatus that separates a high-density culture from a low-density culture medium by a barrier that permits diffusion of solutes (Fig. 18.2). Dialysis tubing works well for this purpose. A high-density cell suspension (the nurse culture) is placed inside the dialysis tubing, which is suspended by a rod in the low-density medium. As the nurse culture

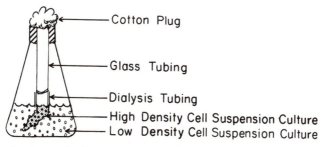

Cotton Plug

Glass Tubing

Dialysis Tubing
High Density Cell Suspension Culture
Low Density Cell Suspension Culture

Fig. 18.2. Apparatus for the separation of a high-density culture from a low-density culture medium.

metabolites are produced and diffuse into the low-density medium, there is a period of maximum conditioning when the growth-promoting activity in the low-density medium rises; the low-density cell suspension should be inoculated during this period. After this point, the growth-promoting activity declines if incubation of the nurse culture is continued, partly because of nutrient depletion of the medium by the nurse culture.

CELL AGGREGATION IN SUSPENSION CULTURES

The degree of cell separation in a suspension culture is greatly influenced by the origin and history of the callus used to initiate the culture and by the composition of the culture medium. Better separation of *Dacus* cells has been achieved when the cells were grown on a fully defined medium supplemented with coconut milk. Vitamin levels higher than those commonly used for parent callus cultures may be required for growth and separation of cells in suspension cultures. Cell separation is often affected by the levels of hormones present in the tissue and/or culture medium. Relatively high auxin levels have been reported to increase cell separation, while low auxin levels decrease separation. The absolute and relative concentrations of both auxins and cytokinins must be considered.

Changes in the extent of cell separation throughout the growth cycle have been reported with *Acer* cultures. Henshaw *et al.* (3) noted that aggregation increases during the period of maximal cell division and the incidence of mitotic index is greater in the cell aggregates than in single cells. Thus, the frequency of subculture may affect the degree of aggregation. Frequent transfers designed to maintain the cells in an active state of cell division may result in increased cell aggregation. For example, the largest number of aggregates in carrots was present during the early growth phase and the cells were more dispersed at the stationary phase (3).

Free cell cultures have been obtained by using enzymes in the culture medium. A low concentration of cell wall-degrading enzymes, plus sorbitol to increase the osmotic potential of the medium, increased cell separation without affecting the growth rate in cultures (3). However, this is a species-dependent phenomenon.

The frequent reports that cell division occurs most frequently in cell aggregates has raised the possibility that high growth rates in cell suspension cultures may depend upon the formation of aggregates. Within these aggregates, there may be a differentiation of the cells into "feeder" cells and/or meristematic cells or the aggregate may achieve a volume to surface ratio compatible with the accumulation of growth-limiting factors.

MEASUREMENTS OF GROWTH

The growth of cell suspension cultures can be followed by measuring selected parameters at intervals during the growth cycle. Fresh weight, dry weight, cell number, mitotic index, packed cell volume, and total protein all are measures of growth. Determination of all except packed cell volume is quite time-consuming.

To determine packed cell volume, small samples (up to 10 ml) are aseptically removed from the suspension culture. The samples should be removed at selected intervals and placed in either glass or plastic graduated, conical centrifuge tubes. The tubes are then centrifuged at 1000 × g for 5 min and the total volume of the packed cells is determined. The packed cell volume is expressed as ml of cell pellet/ml of culture. This technique is only applicable when used with a fine cell suspension. It is also important that consistent times and speeds of centrifugation be used. Another source of error is that the surface of the pellet may not be level after centrifugation and the level may therefore need to be estimated. In addition, cell debris in the sample will be measured by this method even though it does not constitute living cells. The packed cell pellet can be resuspended and macerated to provide samples for determinations of cell number, fresh and dry weights, and other parameters.

A cell count is the most accurate measure of growth of a cell suspension culture. Determination of cell number is a simple but tedious procedure. This first step is to disrupt cell aggregates, which is usually accomplished by treating the suspension with 5–15% chromic acid. A standard procedure for determining cell number consists of the following steps:

1. Add 1 volume of culture to 2 volumes of a 8% chromic acid solution.
2. Heat to 70°C for 2–15 min (depending upon the growth phase harvested).
3. Cool.
4. Shake vigorously for 10 min.
5. Centrifuge the suspension, pour off the chromic acid, and resuspend the pellet in an 8% saline (NaCl) solution for about 10 min.
6. Count the cells using a haemocytometer.

An alternative procedure for breaking up cell aggregates and directions for using a haemocytometer are given in Appendix 9. In some cases, effective cell separation may be obtained by use of a 0.1% pectinase (w/v) solution at pH 3.5.

The fresh weight of a suspension culture is determined by collecting the cells on a preweighed (wet) filter paper. The cells are then washed with water under vacuum and the filter disks reweighed. The difference between the two measurements is the fresh weight of the cells. A large sample is required to obtain accurate results, but otherwise this is a fairly easy method of determining cell growth. Dry weight may also be used as a measure of cell growth. Dry weights are determined in a procedure similar to that for fresh weights, except the filter disks are dried in a forced air oven for 12 hr at 60°C then cooled in a desiccator containing silica gel. Detailed instructions for the determination of dry weights are presented in Appendix 11. Cell fresh and dry weights are normally expressed in grams per milliliter of culture or grams per 10^6 cells.

PLATING OF CELL SUSPENSIONS AND CULTURE OF SINGLE CELLS

The technique of plating cell suspension cultures, first reported by Bergmann in 1960, is of particular importance when attempts are being made to obtain single-cell clones. To plate plant cells, a suspension of cells is mixed with a melted agar medium at a temperature between 30° and 35°C. The cells are dispersed in the medium in such a way that they are evenly distributed and fixed in a thin layer after the agar has cooled and solidified. The objective is to establish as low a cellular density as is compatible with the clone, thus enhancing growth of cellular clones from single cells or small aggregates.

A cell suspension to be plated should be freed from large aggregates by aseptic filtration, so that the filtrate contains many single cells or small aggregates containing 5–10 cells, which are likely to have been derived from single cells. The removal of large aggregates is important for two reasons: (1) large aggregates undergo cell division at a higher rate than do small aggregates or single-cell units and (2) large aggregates begin cell division sooner after transfer than do single cells or small aggregates (1).

The basic technique for plating cells involves the following steps:

1. The cell suspension is filtered so that any large aggregates are removed from the suspension.
2. The cell number of the filtrate is determined. Based on this count, a known number of cellular units can be established per unit volume of plating medium.
3. The counted suspension is adjusted by dilution or by concentration (through low speed centrifugation and removal of supernatant) so that a 2-ml suspension when inoculated will give the required cellular unit density.
4. Sterile culture medium containing 0.6% agar is prepared and cooled to 35°C.
5. 10 ml of the medium and 2 ml of cell suspension are mixed and poured into a sterile disposable petri dish.
6. The dish is swirled to ensure that the cells are properly mixed, sealed with Parafilm, and incubated.

The cellular units (aggregates or cells) are then examined directly through the plate cover, and cell number is determined by counting the number of units per known area using a 40× magnifying stereomicroscope. At least 20 fields of known area should be examined, and the number of cellular units/mm^2 and per plate calculated. Following incubation, 20 random grids (fields) of known area are counted using a stereomicroscope to determine the number of visible colonies per unit area or per plate. This value can be used to calculate the plating efficiency (PE), as follows:

$$PE = \frac{\text{Final number of colonies/plate}}{\text{Initial number of cellular units/plate}} \times 100$$

A shadowgraph print may be made of the plates to help in counting. This is done by placing a plate on top of a photographic document paper (e.g., DR Ilfoprint DR 3 5L No. 3 Projection Document Matt Lightweight), under a light source such as a photographic enlarger. A negative of a ruled area is placed in the negative holder of the enlarger, and the paper is exposed for the desired time. The result will be a print of the petri plate on which the grid has been superimposed.

Effects of Population Density

In plating single cells from higher plants, it is usually necessary to use cell densities of 10^3–10^5 cells/ml to obtain a high plating efficiency. The cell density required for cell division and colony formation may be reduced by using a conditioned medium. With tobacco cultures (6), to achieve a plating efficiency of 50%, a density of 360 cells/ml was required when plated on a dish with 6300 mm^2; to achieve an efficiency approaching 100% the density had to be raised to 720 cells/ml (Table 18.1).

To obtain the highest plating efficiencies, several conditions should be observed. First, either a conditioned medium or a synthetic medium especially designed to permit growth from a low initial cell density should be used. Cells held too long in the stationary phase should not be used for plating; cells harvested during the exponential growth phase show the highest plating efficiency. In addition, the cells should never be exposed to temperatures higher than 35°C during the plating procedure, and the plates should be incubated in low light or the dark.

Growth of Cells in Microchambers

De Ropp (2) was the first to attempt culturing single cells in hanging drops of medium. He found that only cell aggregates of 10 cells or more would divide. Torrey (10), using a double coverslip technique, induced division of single cells from pea root callus that was located around a nurse callus. Approximately 8% of the cells remained alive long enough to begin division. The largest aggregate from a single cell contained only seven cells before the culture died.

The best success in growing single cells in microchambers has been achieved by Jones et al. (4). Using single cells derived from hybrids of

Table 18.1.   Effect of Population Density on Plating Efficiency of Dark-grown Cells of *N. Tabacum*

Cells/ml	Cells/Plate	Planting Efficiency	Colonies/mm^2
90	1350	0	0
180	2700	9.9 ± 3.1	0.04
360	5400	45.7 ± 6.1	0.4
720	10,800	90–100	1.7

Source: Logemann and Bergman (6).

Nicotiniana tobacum × *N. glutinosa*, they obtained at least 30 cells in the chamber when a conditioned medium was used.

REFERENCES

1. Blakely, L. M. and F. C. Steward. 1962. The growth of free cells, II. Observations on individual cells and their subsequent patterns of growth, III. The observation and isolation of variant strains. Am. J. Bot. 49:653.
2. DeRopp, R. S. 1955. The growth and behavior *in vitro* of isolated plant cells. Proc. Roy. Soc. Lond. B. 144:86–93.
3. Henshaw, G. G., K. K. Jha, A. R. Mehta, D. J. Shakeshaft, and H. E. Street. 1966. Studies on the growth in culture of plant cells I. Growth patterns in batch propagated suspension cultures. J. Exp. Bot. 17:362–377.
4. Jones, L. E., A. C. Hilderbrandt, A. J. Riker, and J. H. Wu. 1960. Growth of somatic tobacco cells in microculture. Am. J. Bot. 47:468–475.
5. Lamport, D. T. A. 1964. Cell suspension cultures of higher plants: isolation and growth energetics. Exp. Cell Res. 33:195–206.
6. Logemann, H. and L. Bergman. 1974. EinfulB vonLight und medium auf die "Plating Efficiency" isolierter zellen aus Calluskulturen von *Nicotiana tabacum* var. "Samsun". Planta 121:283–292.
7. Steward F. C., S. M. Caplin, and F. K. Millar. 1952. Investigation of growth and metabolism of plant cells. I. New techniques for the investigation of metabolism, nutrition and growth in undifferentiated cells. Ann. Bot. 16:58–77.
8. Steward, F. C. and E. M. Shantz. 1965. The chemical induction of growth in plant tissue cultures. pp. 165–186. *In* The Chemistry and Mode of Action of Plant Growth Substances. R. L. Wain and F. Wightman (editors). Butterworths Ltd., London.
9. Stuart, R. and H. E. Street. 1969. Studies on the growth in culture of plant cells. IV. The initiation and division in suspensions of stationary phase cells of *Acer pseudoplatanus*. J. Exp. Bot. 20:556–571.
10. Torrey, J. G. and J. Reinert. 1961. Suspension culture of higher plant cells in synthetic medium. Plt. Physiol., Lancaster 36:483–491.

19

Establishment of Carrot Cell Suspension Cultures

The culture of plant tissues in an agitated liquid medium eliminates many of the disadvantages associated with the culture of tissue on agar. Movement of the tissue in relation to the nutrient medium facilitates gaseous exchange, removes any polarity of the tissue due to gravity, and eliminates nutrient gradients within the medium and at the surface of the cells. The incubation of a friable callus in a liquid nutrient medium agitated on a gyratory shaker will eventually give a suspension of cells that is more amenable to experimental manipulation than callus grown on agar. A cell suspension should be manipulated so that it consists of single cells and very small aggregates of 2–15 cells. Both the composition of the medium and the frequency of subculture influence the degree of aggregation.

The growth pattern of suspension cultures depends upon the cell density per ml of medium at inoculation. With a suboptimal inoculum density, growth is either slow or does not occur at all; with a supraoptimal inoculum density, the lag phase is reduced but so is the growth rate, and growth ceases early. The exact relationship between inoculum size and growth pattern varies among species. For carrot cell suspensions subcultured at 7-day intervals, the optimal inoculum size is $1-3 \times 10^5$ cells/ml or 100–300 mg (fresh weight) in a volume of 60 ml of fresh medium (1).

Materials Required
1. 125-ml Erlenmeyer flasks for cultures
2. 3 spatulas
3. 25 sheets aluminum foil (100 × 100 mm), sterile
4. 100-ml graduated cylinders, sterile, sealed with aluminum foil
5. 2 nylon or stainless steel sieves with a porosity of 250 μM
6. 5 sterile petri dishes
7. Waterproof marking pen
8. Rotary shaker with adjustable speed between 80–100 rpm

9. Bunsen or ethanol burner
10. 150-ml Erlenmeyer flask containing 100 ml of 95% ethanol
11. Roll of Parafilm
12. Carrot medium

MS basal salts[1]	
Sucrose	20.0 g/liter
Nicotinic Acid	0.5 mg/liter
Pyridoxine	0.1 mg/liter
Thiamin	1.0 mg/liter
Glycine	3.0 mg/liter
Myo-inositol	100.0 mg/liter
2,4-D	4.5 μM

13. Friable carrot (*Daucus carota*) callus cultures, 7–10 days after subculture

Procedures

1. Remove culture tubes containing carrot callus from the rack and empty the callus into a sterile petri plate. Using a spatula, transfer 3 or 4 pieces of callus (ca. 1-g each) into a 125-ml Erlenmeyer flask containing 40 ml of medium. Flame the neck of the flask before and after inoculating with callus; seal the flask with aluminum foil and secure the foil cap with Parafilm. Repeat this operation until all flasks have been inoculated. Place the flasks on a rotary shaker in the dark at 25°C.
2. After 7 days, remove all flasks from the shaker and transfer to a sterile hood.
3. Flame the neck of each flask, then pour the contents through a sieve into a sterile 100-ml graduated cylinder. Recap the cylinder.
4. Allow the contents in the graduated cylinder to settle for approximately 10 min and then pour off the supernatant. Empty the residual cells in the graduated cylinder into a clean 125-ml flask containing 40 ml of fresh medium, flame the neck, recap with foil, and incubate as described in step 1.
5. Subculture again after 7 days as described in steps 2–4, using only one-fifth of the residual cells as the inoculum.
6. After three or four culture periods, carrot cell suspensions can be subcultured by transferring 10 ml of the cell suspension with a wide-bored 10-ml pipette to 50 ml of fresh medium.
7. When preparing to make transfers in cell cultures, at least one sample from the old culture should be taken and submitted to a

[1] Prepare as described in Chapter 2 or from prepackaged salts.

cell count to determine cell population density. This will be required to calculate and establish the proper cell density in the new culture. A repersenatice sample can be taken using a pipette or syringe with a long needle.

Scheduling

Event	Timing
Initiation of cell suspension cultures and determination of initial total cell number and of packed cell volume	Day 0
Determination of total cell number and packed cell volume	Day 1, 2, 4, and 7
First transfer of cultures	Day 7
Subsequent transfers of cultures	Day 14, 21, 28, etc.

Recording Results
1. Record all details of setting up the experiments.
2. Make visual observations of the suspension cultures at weekly intervals recording changes in the morphology.
3. Determine the total cell number and the packed cell volume at the beginning and end of each subculture period. Refer to Chapter 18.
4. After three subcultures, estimate the total cell number and packed cell volume initially and again at days 1, 2, 4, and 7 and plot these points on a graph.

REFERENCES

1. Henshaw, G. G., K. K. Iha, A. R. Mehta, D. J. Shakeshaft, and H. E. Street. 1966. Studies on the growth in culture of plant cells. I. Growth patterns in batch-propagated suspension cultures. J. Exp. Bot. 17:362–377.
2. Reinert, J. and M. M. Yeoman. 1982. Plant Cell and Tissue Culture—A Laboratory Manual. Springer-Verlag, New York.
3. Street, H. E. 1977. Cell (suspension) culture techniques. pp. 61–102. *In* Plant Tissue and Cell Culture. H. E. Street (editor). Botanical Monographs 11. Blackwell Publications, Oxford.
4. Torrey, J. G. and J. Reinert. 1961. Suspension cultures of higher plant cells in synthetic media. Plant Physiol. 36:483–491.

20

Plating of Cell Suspension Cultures

As discussed in Chapter 18, the technique of plating plant cells involves mixing a cell suspension with melted agar medium (30°–35°C) and then dispersing the mixture so that the cells become evenly distributed and fixed in a thin layer after the medium has cooled and solidified. If the proper culture medium is chosen and the cells are properly handled, small colonies of single-cell origin will develop from plated cells.

To obtain high plating efficiencies, it is usually necessary to plate at densities of 10^3 to 10^5 cells/ml. The cell density needed to induce cell division and colony formation can be reduced by using a conditioned medium or medium supplemented with yeast extract, coconut milk, or other organic supplements. The recognition of the need for a minimum inoculum density for the successful growth of both callus and suspension cultures and the exposure effect of both callus tissue masses and high-density cell suspensions on single cells point to a mutually beneficial interaction between cultured cells and a dependence of each cell upon the cell population of which it is a constituent.

In a low population density situation growth will be initiated only if the basic medium is:

1. Supplemented by conditioning or by the addition of the required metabolites at appropriate concentrations.
2. Altered in some way which:
    a. enhances the rate of synthesis by the cells of these metabolites.
    b. reduces the efflux of these metabolites from these cells (Synthetic conditioned medium).

Enhanced plating efficiencies have been achieved by using a standard basic medium supplemented with selected growth factors, such as cytokinins and amino acids. In cell suspension cultures, the minimal effective initial cell density is $9-15 \times 10^3$ cells/ml with a standard culture medium. This required cell density can be reduced to 2×10^3 cells/ml by use of a synthetically conditioned (i.e., supplemented) medium and to $1.0-1.5 \times 10^3$ cells/ml by use of a conditioned medium.

Single isolated cells provide excellent experimental systems for studies of cloning and genetic stability, cellular growth requirements, host–pathogen interactions, and many other phenomena. However, to fully exploit the genetic variability expressed by individual cells, one must be able to plate cells and induce subsequent growth in a large percentage of the plated cells.

Materials Required
1. Sterile plastic petri dishes (15×100 mm)
2. Hot plate
3. Glass rod (200×2 mm) and 3 spatulas
4. Roll of Parafilm
5. Bunsen or alcohol burner
6. Waterproof marking pen
7. 250-ml beaker containing 150 ml of 95% ethanol
8. 500-ml beaker filled with 250 ml of water
9. A. Basic culture medium prepared as follows:

MS macro- and micronutrients[1]		
Thiamin·HCl	1.0	mg/liter
Myo-inositol	100.0	mg/liter
$NaH_2PO_4 \cdot 2H_2O$	1.65	g/liter
Sucrose	10.0	g/liter
2,4-D	0.5	mg/liter

   B. Solid medium for plating cells

Basic culture media as described in 9A		
Agar	8.0	g/liter

10. Stock carrot (*Daucus carota*) callus cultures subcultured on standard medium described above within 10 days of initiation of experiment
11. Actively growing carrot cell suspension cultures subcultured into standard medium (without agar) about 3 days before initiation of experiment
12. Compound microscope for determining cell number
13. 10 sterile large orifice pipettes (10 ml)
14. 10 sterile large orifice pipettes (1 ml)
15. Inverted phase contrast microscope

[1] Prepared as in Chapter 2 or from prepackaged salts.

Procedures

A-1. Determine the cell number in cell suspension cultures as described in Chapter 18. By dilution with culture medium as described in 9, prepare aliquots containing 100, 1000, and 10,000 cells/ml. You will need about 10 ml of each cell population for subsequent plating.

A-2. Dispense 10 ml of standard culture medium into each of 12 petri plates. Allow 6 of the plates to solidify.

A-3. Allow the medium in 6 plates to cool to 30°–35°C and then add 1.0 ml of cell suspension to each plate. Inoculate 2 plates with each of the cell populations prepared in step A-1. Gently swirl each plate to ensure proper mixing of the cells within the agar medium. Seal the plates with Parafilm, label with medium and inoculum density, and incubate in the dark at 25°C.

A-4. Inoculate each of the 6 plates containing solidified medium with 1.0 ml of cell suspension, preparing 2 replicates with each of the cell populations. Spread the cell suspension on the surface of the agar with a glass rod as illustrated in Fig. 20.1. Take care not to damage the solidified agar. Seal all plates with Parafilm, label them, and incubate in the dark at 25°C.

Plating on Callus-Conditioned Medium

B-1. Obtain 6 tubes containing stock carrot callus cultures. Carefully remove the callus from each tube with a spatula and place the callus in a sterile plastic petri plate. (Save callus for later use.) Recap the tube and place it in a boiling water bath. As soon as the agar melts, remove the culture tube from the boiling water; under sterile conditions, remove the cap, flame the open end and pour the melted media into two petri plates, ca. 10 ml of media per plate. Repeat this procedure for all callus culture tubes. You will have 12 plates in all. Allow 6 of the plates to solidify.

B-2. Plate out cell suspensions as described in steps A-3 and A-4, using the three cell populations (100, 1000, and 10,000 cells/ml) prepared in step A-1.

Fig. 20.1. Technique for spreading cell suspension inoculum on surface of solidified medium.

Plating on Nurse Culture-Conditioned Medium
C-1. Dispense 10 ml of standard culture medium into each of 12 petri plates.
C-2. Prepare 12 plates with each containing ca. 10 ml of callus-conditioned medium as described in step B-1.
C-3. Select 6 plates from steps C-1 and C-2 before the medium has solidified. Using the procedure described in step A-3, inoculate 2 plates from each group with each of the cell populations prepared in step A-1. Do *not* seal the plates.
C-4. Allow the remaining 6 plates from each group prepared in steps C-1 and C-2 to solidify. Using the procedure described in step A-4, inoculate 2 plates from each group with each of the cell populations. Do *not* seal the plates.
C-5. Cut the callus tissue obtained in step B-1 into ca. 50-mg sections. Transfer 4 pieces of callus to each of the petri plates prepared in this section (a total of 24 plates). Seal all plates with Parafilm, label, and incubate them in the dark at 25°C.

Observing Growth of Single Cells
1. Immediately following inoculation of all plates, locate 10 single cells with the aid of an inverted phase microscope on the agar layer of each plate and draw a circle around the cell.
2. Check the growth of these single cells every other day for the next 2 weeks.
3. After 28 days, count the number of colonies in 20 random fields of known area on each plate. Calculate the number of colonies/mm^2 and the number of colonies/plate.

Scheduling

Event	Timing
Initiation of plated cultures	Day 0
Begin observation of isolated single cells	Day 2
End visual microscopic observation	Day 14
Terminate experiment and determine the number of colonies formed	Day 28

Recording Results
1. Record all details of setting up the experiment.
2. Record visual observations of single cell development.
3. Record observations on the effect of inoculum density, culture medium, and plating technique on the number of colonies formed.

REFERENCES

1. Bergman, L. 1977. Plating of plant cells. pp. 213–225. *In* Plant Tissue Culture and its Biotechnological Application. W. Bavy, E. Reinhard, and M. H. Zenk (editors). Springer-Verlag, New York.

2. Hilderbrandt, A. C. 1977. Single cell culture, protoplasts and plant viruses. pp. 581–597. *In* Applied and Fundamental Aspects of Plant Cell Tissue and Organ Culture. J. Reinert and Y. P. S. Bajaj (editors). Springer-Verlag, New York.

3. Reinert, J. and M. M. Yeoman. 1982. Plant Cell and Tissue Culture. Springer-Verlag, New York.

4. Torrey, J. G. 1957. Cell division in isolated single cells *in vitro*. Proc. Nat. Acad. Sci. 43:887–891.

21

Embryogenesis in Carrot Cell Suspension Cultures

Under controlled growth conditions, a callus mass is capable of initiating organ primordia. Another type of organized development is embryogenesis, which is commonly observed in cell suspension cultures. The embryos formed in *in vitro* cultures are capable of forming complete plantlets passing through stages similar to those that occur in normal embryogeny.

The decisive feature of an embryo is its bipolarity, characterized by a shoot and a root pole at opposite ends. When embryos are formed in callus cultures, diffusion gradients in the tissue may determine polarity. The stimulus that triggers polarity in cell suspension cultures, however, is more difficult to explain. Konar et al. (2) reported that potential embryoidal cells were differentiated from cellular aggregates. Isolated carrot cells have shown the capacity to produce embryos, and Reinert et al. (6) demonstrated through microscopic observation that a multicellular aggregate was formed before embryo formation in carrot cultures. The formation of aggregates in such embryoidal-like cells in carrot differed from the typical pattern of embryogenesis of an egg cell in the early stages but had similar stages of ontogeny later on (3).

The nutrient factors that initiate embryoid formation have been widely studied. Reinert (4) reported that the transfer of tissue from an auxin-containing medium to an auxin-free medium and/or the addition of nitrogenous compounds initiated embryogenesis in carrot cultures. The addition of inorganic nitrogen (KNO_3 or NH_4NO_3) or organic nitrogen (amino acids and amides) can stimulate initiation of embryogenesis. Embryogenesis may depend not only on the nitrogen concentration but also on the source of nitrogen. Embryos have never been formed with nitrate as the sole source of nitrogen at any physiological concentration. However, nitrate may be required for the development of later stages of embryos (i.e., heart or globular). Reinert (5) postulated that the conversion of a somatic cell to an embryo may be regulated not solely by nitrogen source and concentration but rather by the interaction of nitrogen source and concentration with auxin concentration.

Materials Required
1. 40 sterile petri plates (15 × 100 mm)
2. 125-ml Erlenmeyer flasks with closures for cultures
3. Waterproof marking pen and stick-on labels
4. Bunsen or ethanol burner
5. 3 spatulas and scalpels
6. 250-ml beaker with 200 ml of 95% ethanol
7. Rotary shaker set at 160 rpm
8. Culture medium containing:

    A. Suspension culture medium
       MS macro- and micronutrients[1]

Sucrose	30.0 g/liter
Thiamin · HCl	1.0 mg/liter
Myo-inositol	100.0 mg/liter
2,4-D	4.5 μM (1.0 mg/l)

    B. Embryo production medium
       MS macro- and micronutrients[1]

Sucrose	30.0 g/liter
Thiamin	1.0 mg/liter
Myo-inositol	100.0 mg/liter

    C. Plating medium
       1/2 MS macro- and micronutrients[1]

Sucrose	30.0 g/liter
Thiamin · HCl	1.0 mg/liter
Myo-inositol	100.0 mg/liter
Agar	8.0 g/liter

9. Actively growing carrot (*Daucus carota*) callus cultures

Procedures
1. Prepare suspension culture medium as described in previous section, dispense into 125-ml Erlenmeyer flasks (50 ml/flask), and sterilize.
2. Place callus from stock cultures into a sterile petri dish and then transfer about 2.5–3.5 g into a culture flask containing suspension culture medium. Repeat until all flasks are inoculated. Place flasks on rotary shaker set at 160 rpm and maintain at 25°C illuminated at 1000 lux (Fig. 21.1C).

---

[1] Prepared as described in Chapter 2 or from prepackaged salts.

Planting of embryos

10-18 days after transfer to embryogenic induction medium.

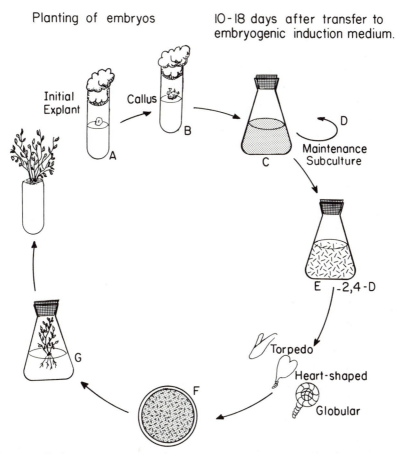

Fig. 21.1. Embryogenesis in carrot starting from initial explant and initiation of callus culture (A). After callus has been subcultured (B) and is well developed, tissue is transferred to cell suspension culture medium containing 2,4-D (C) and is subcultured several times (D). Cells are transferred to embryo induction medium (E) without 2,4-D, developing embryos are plated out (F). Plantlets may be transferred to soil mixture for further development.

3. Subculture the suspension cultures at 14-day intervals onto fresh identical medium. This is accomplished by pouring the suspension into a 50-ml centrifuge tube and spinning at 1000 × g for 5 min. After centrifugation, decant or siphon off all the old media and resuspend the pellet in 50 ml of fresh medium. Cell cultures will generally need to be subdivided every other transfer period. This is accomplished by centrifugation of the suspension as described above, removing the old medium, then resuspending the

pellet in 10–20 ml of fresh medium. Five to ten ml of the resuspended pellet is then pipetted into 50 ml of fresh medium (Fig. 21.1D).

4. Prepare embryo initiation medium as described in previous section, dispense into 125-ml Erlenmeyer flasks (50 ml/flask), and sterilize.

5. After 2–4 subcultures on the suspension culture medium (step 2), decant the medium and transfer the cells as described in step 3 to a flask containing embryo initiation medium (Fig. 21.1E). Incubate the cultures at 25°C at 1000 lux.

6. Prepare and sterilize plating medium as described in previous section. Dispense 10 ml/plate into sterile petri plates.

7. After cells have grown for 10–18 days in the embryo induction medium, remove a 5 ml-aliquot and plate out on plates prepared in step 6 before the medium has solidified (Fig. 21.1F). Refer to the exercise in Chapter 20 (step A-3) for a description of the plating out procedure. Seal the plates with Parafilm and incubate at 25°C illuminated at 1000 lux.

8. Plantlets can be transferred after 30–60 days to peat pots or vermiculite for further development (Fig. 21.1G).

Scheduling

Event	Timing
Initiation of cell suspension culture	Day 0
Subsequent subcultures on cell suspension medium	Every 10–18 days
Transfer to embryo induction medium	Day 40–as needed
Plating of cells for plantlet development	30–60 days after inoculation onto embryo induction medium

Recording Results
1. Record all details of setting up this experiment.
2. Record date of all transfers
3. Observe with a stereomicroscope embryo formation every day and draw sketches of the embryonic structures observed.

REFERENCES

1. Evans, P. A., W. R. Sharp, and C. E. Flick. 1981. Growth and behavior of cell cultures: Embryogenesis and organogenesis. pp. 45–113. *In* Plant Tissue Culture:

Methods and Applications in Agriculture. T. A. Thorpe (editor). Academic Press, New York.

2. Konar, R. N., E. Thomas, and H. E. Street. 1972. Origin and structure of embryoids arising from epidermal cells of the stem of *Ranunculus sceleratus* L. J. Cell Sci. 11:77–93.

3. McWilliam, A. A., S. M. Smith, and H. E. Street. 1974. The origin and development of embryoids in suspension cultures of carrot (*Daucus carota*). Ann. Bot. 38:243–250.

4. Reinert, J. 1959. Uber die Kontrolle der Morphogenese Und die Induktion von Advenivembryonen and Gewebekulturen aus Karotten. Planta 53:318–333.

5. Reinert, J. 1973. Aspects of organization—Organogenesis and embryogenesis. pp. 338–355. *In* Plant Tissue and Cell Culture. H. E. Street (editor). Blackwell Publications, Oxford.

6. Reinert, J., D. Backs-Husemann, and H. Zerban. 1971. Determination of embryo and root formation in tissue cultures from *Daucus carota*. *In* Les Cultures des Tissue de Plantes. Centre Natl. Rech. Sci., Paris.

22

Isolation and Culture of Single Cells

The nurse technique, petri-dish plating, and growth of isolated cells in microchambers are all methods which may be used to culture isolated single cells. Studies of the requirements for growth of isolated single cells not only advance our knowledge of cell physiology but are essential if the full extent of variation in cultured cells is to be exposed. The recognition of the need for a minimum inoculum density for the successful initiation of both callus and suspension cultures and the exposure effect of both callus tissue masses and high density cell suspensions on single cells points to a mutually beneficial interaction between cultured cells and a dependence of each cell upon the cell population of which it is a constituent.

The early attempt by De Ropp (1) to culture single cells of plant tissues by the hanging drop method was unsuccessful; only cellular aggregates of at least 10 cells showed any mitosis. Torrey (5) used the Maximov double coverslip method (Fig. 22.1) to induce division of single cells derived from pear root callus. Approximately 8% of the cells that were cultured in this manner began to divide; however, cell division did not continue very long, as the largest colony contained only 7 cells. The most successful work on the culture of single cells has been reported by Jones et al. (2), who used microchambers made from coverslips and mineral oil. They found that division occurred when 30 or more cells were present in the chamber.

Single-cell clones isolated from established callus and cell suspension cultures have exhibited the heterogeneity of the parent cultures; that is, the single-cell clones from a parent tissue show variation in color, texture, morphogenetic potential, growth rate, etc. The use of single-cell cultures can help advance knowledge of cell physiology and exploit to the full extent the genetic variation present in callus and cell cultures.

SPECIAL GROWTH REQUIREMENTS OF SINGLE CELLS

The nurse culture system shows that the nurse callus supplies by diffusion through a paper barrier not only those essential nutrients supplied by the culture medium but also those extra metabolites from the callus required for induction of division. Similar results are likewise demonstrated when a callus mass is implanted on a petri dish seeded with isolated single cells. In the latter case, cell division begins first in the region adjacent to the nurse callus then the front moves outward from the nurse callus. Similar results may also be observed when single cells which form colonies first serve as a nurse callus for subsequent isolated cells.

The fact that a medium which has supported growth of a callus mass of liquid suspension can now support growth of cultures below their critical initial cell density, indicates that the medium has become "conditioned" by the release of beneficial metabolites in the culture medium. The success of the micro-chamber described by Jones et al. (2) can be interpreted as a consequence of the individual cell being capable of conditioning the small volume of medium which it is suspended in.

A concept which emerges from these considerations is that the minimal or basic medium for the growth of cells from a high population density is simpler than that for the growth from a low population density which is simpler than that for the growth of a single cell. This is postulated to be due to the "leaky" nature of the cultured cells. Each cell is presumably capable of synthesizing all the metabolites essential for growth and division but the required endogenous concentrations of certain critical metabolites have to be established under conditions where they are continuously being lost to the bathing medium. Therefore, only when the concentrations of these metabolites reach an appropriate level *external* to the cell, can cell division begin. In other words, an equilibrium between the internal and external concentration of the metabolites must be reached prior to the induction of division. The situation occurs rapidly in a high population density but may never occur if a single cell is in a infinite volume of medium. Therefore, in a low population density situation growth will be initiated only if the basic medium is:

1. supplemented by conditioning the medium or by the addition of the required metabolites at appropriate concentrations.
2. altered in some way which:

   a. enhances the rate of synthesis by the cells of these metabolites, or

b.   reduces the efflux of these metabolites from the cells. [Such a modified media has been called a "Synthetic conditioned medium"—Stuart and Street (4)].

Such a hypothesis is in agreement with observations where division occurs first and with greater frequency in cell aggregates than in free cells of a plated suspension. The observation that adjusting the pH of the culture medium and raising the level of iron suggests that factors affecting cell permeability and the ionic status of the tissue may be important in conditioning (4).

Wood and Brown (6) demonstrated that the addition of KCl, $NaNO_3$, NaH_2PO_4 and $(NH_4)_2SO_4$ to a basic White's medium would not only enhance the level of cell growth but would induce cell growth in a medium void of auxin. Therefore, increased plating efficiencies may be obtained by paying more attention to the inorganic ionic composition of the plating medium.

Enhanced planting efficiencies have been achieved by using a standard basic medium supplemented with selected metabolites such as cytokinins and amino acids. By using a synthetically conditioned medium, the minimal effective initial cell density may be lowered as follows:

Standard medium	$9–15 \times 10^3$ cells/ml
Synthetically conditioned	2×10^3 cells/ml
Conditioned	$1.0–1.5 \times 10^3$ cells/ml

There is now evidence that volatile factors may also play a role in the conditioning of the medium. Stuart and Street (4) showed that the initial cell density could be dropped to 600 cells/ml when exposed to the culture atmosphere of an actively growing medium. This work showed that when CO_2 was omitted from the culture medium, growth at low densities was inhibited. Subsequent work with low density plating efficiencies does show similar results (see Table 22.1). Concentra-

Table 22.1.   Plating Efficiency in Presence and Absence of CO_2 of *Acer pseudoplatinus* Plated at Low Densities

Initial Cell Densities (ml/l)	No CO_2	1% CO_2
500	0%	110%
1000	42%	95%
2000	60%	101%

Source: Street (3).

tions of CO_2 up to 1–1.5% may stimulate growth but above 2% it inhibits growth.

Materials Required
1. Laminar flow hood
2. Phase-contrast microscope
3. Binocular dissecting microscope (40 × preferred)
4. Sterile microscope slides
5. Sterile 22 × 22 mm coverslips
6. Sterile 22 × 40 mm coverslips
7. Sterile mineral oil (335/350 viscosity)
8. Sterile glass or plastic petri plates
9. Forceps
10. Bunsen or alcohol burner
11. Sterile finely drawn Pasteur pipettes
12. Sterile medium finely drawn Pasteur pipettes
13. Pipette bulbs
14. Well-type microplate with 96 wells
15. Roll of Parafilm
16. 250 ml 95% ethanol
17. Liquid MS medium supplemented with 30 g/liter sucrose, 0.4 mg/liter thiamin HCl, 0.1 mg/liter nicotinic acid, 0.1 mg/liter pyridoxine HCl, 5.0 mg/liter 2,4-D, and 1.0 mg/liter BA
18. Carrot or creeping bentgrass cell suspension cultures

Preparation of Microculture Slides
1. Place a drop of mineral oil about 1 cm from the right end of a microscope slide and another about 3 cm from it using aseptic techniques (Fig. 22.1A).
2. Gently lower a 22 × 22 mm coverslip over each drop of oil (Fig. 22.1B).
3. Place a ring of sterile mineral oil between the coverslips and three small drops of oil on each coverslip (Fig. 22.1C).
4. Pour an aliquot of the cell suspension into a sterile petri plate. Using the binocular microscope, find a single cell or filament in the suspension. With a Pasteur pipette, remove a cell from the suspension. The following tricks will help you in securing cells: (1) use a finely drawn pipette, (2) fill the pipette slightly before placing the tip in the suspension, and (3) keep a steady hand and light control on the pipette bulb.
5. Place the selected cell or cells on the microscope slide in the center of the mineral oil ring (Fig. 22.1D). Then place a drop of liquid MS medium in the ring and carefully place a 22 × 40 mm

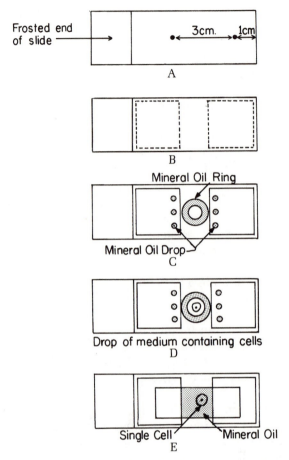

Fig. 22.1. Preparation of microculture slides for the culture of single cells using the double coverslip method.

coverslip over the ring containing the cell and medium (Fig. 22.1E), forming a small microchamber. Label the frosted end of the slide with the slide number, cell type, date, etc.

6. Repeat steps 1–5 until 12 microplates have been prepared.
7. Observe the cell(s) in the microchambers using a phase contrast microscope. Look for various cell components such as the nucleus, nucleolus, cytoplasmic streaming, mitochondria, and plastids.
8. Store each microchamber in a petri plate with several drops of media placed around the slide. Seal the plate with Parafilm and incubate in the dark at 25°C.

9. Check slides daily to observe cell growth and watch for cell division. Determine cell number in each chamber after 2 weeks.

Preparation of Microplates
1. Supplement the MS medium with the following eight treatments:

Glycine	0.0 mg/liter
Glycine	2.0 mg/liter
Glycine	4.0 mg/liter
Glycine	6.0 mg/liter
Tyrosine	0.0 mg/liter
Tyrosine	2.0 mg/liter
Tyrosine	4.0 mg/liter
Tyrosine	6.0 mg/liter

2. Place 6–8 drops of each culture medium treatment into one of the wells in a microplate. Label each treatment.
3. Remove a cell(s) from the cell suspension as described in step 4 of the previous section and place it in one of the treatment wells of the microplate. Cover the medium with 2–3 drops of sterile mineral oil. This will prevent evaporation and make observation easier. Continue to transfer cells until all 8 treatment wells are inoculated.
4. Repeat steps 2 and 3 until all 96 wells have been inoculated.
5. Cover the plate and seal the edges with Parafilm. Incubate in the dark at 25°C.
6. Observe cells with a phase contrast microscope. Check microplates daily for cell growth and division. Determine cell number after 2 weeks.

Scheduling

Event	Timing
Initiation of culture	Day 0
Visual check of the culture	Daily basis
Addition of fresh medium	Day 7, 14, etc.

Recording Results
1. Record all details of setting up the experiment.
2. Record visual observations of the cell growth on a daily basis. Indicate when division of the cells begins and how frequently division occurs.
3. Record when the culture media is added and how much is added at each interval.

REFERENCES

1. De Ropp, R. S. 1955. The growth and behaviour *in vitro* of isolated plant cells. Proc. Roy. Soc. Lond. 144:86–93.
2. Jones, L. E., A. C. Hilderbrandt, A. J. Riker, and J. H. W. Wa. 1960. Growth of somatic tobacco cells in microculture. Am. J. Bot. 47:468–475.
3. Street, H. E. 1973. Single-cell clones. pp. 191–204. *In* Plant Tissue and Cell Culture. H. E. Street (editor). Blackwell Publication, Oxford.
4. Stuart, R. and H. E. Street. 1971. Studies on the growth in culture of plant cells. X. Further studies on the conditioning of culture media by suspensions of *Acer pseudoplatanus* L. J. Exp. Bot. 22:96–106.
5. Torrey, J. G. 1957. Cell division in isolated single plant cells *in vitro*. Proc. Nat. Acad. Sci. 43:887–891.
6. Wood, H. N. and A. C. Braun. 1961. Studies on the regulation of certain essential biosynthetic systems in normal and crown-gall tumor cells. Proc. Nat. Acad. Sci. (USA) 67:1283–1287.

23

Potato Micropropagation and Germplasm Storage

A number of different approaches have been successfully used in the regeneration of potato plants from *in vitro* cultures. Mellor and Stace-Smith (6) rooted excised bulbs on a filter-paper wick culture using a MS medium with no exogenous hormones. Lam (5) induced embryoids from tuber disks cultured on a medium of MS salts and Nitsch and Nitsch organics supplemented with 0.4 mg/liter IAA, 0.4 mg/liter GA$_3$, 0.8 mg/liter kinetin, and 1.0 mg/liter casein. Shoots developed from the embryoids when they were transferred to a similar medium containing only 0.4 mg/liter BA. This procedure however resulted in the formation of many "abnormal shoots" which were reddish in color and had unexpanded leafy structures. Skirvin *et al.* (9) reported that when these abnormal shoots were cultured on a Whites medium with no growth regulators, normal shoot development would begin to occur within 3–4 days and normal plantlets could be transferred to the soil within 3–4 weeks. Roest and Bokelman (7) obtained plantlet regeneration from potato stem segments when explants were cultured on a MS medium supplemented with 10 mg/liter GA$_3$, 1.0 mg/liter BA, and 1.0 mg/liter IAA.

Espinoza *et al.* (3) have reported on the micropropagation of potato by either nodal section or shake cultures. They found that when nodal sections were inoculated onto a MS culture medium supplemented with 0.25 mg/liter GA$_3$ and 2.0 mg/liter calcium pantothenic acid, the number of nodes increased sixfold within 3–4 weeks. When nodal sections were cultured on a liquid MS medium supplemented with 0.4 mg/liter GA$_3$, 0.5 mg/liter BA, 0.01 mg/liter NAA, 2.0 mg/liter calcium pantothenic acid, and 2% sucrose, there was a 10- to 20-fold increase in the number of nodes in 2–3 weeks.

Haploid and dihaploid potato plants have been produced using anther culture. Dunwell and Sunderland (2) noted that embryoid development from some of the callus derived from anthers cultured on a MS medium supplemented with 0.01–5.0 mg/liter kinetin at NAA concentrations above 1.0 mg/liter. Foroughi-Wehr *et al.* (4) induced monohaploid plants to develop from anthers cultured on a LS medium containing 1.0 mg/liter IAA and 1.0 mg/liter BA.

Plant regeneration from potato protoplasts has also been reported. Shepard and Totten (8) were able to regenerate callus and then induce shoot and root formation from potato mesophyll protoplasts. They used the medium described by Lam (5) supplemented with 0.1 mg/liter IAA, 0.5 mg/liter zeatin, and 40.0 mg/liter adenine sulfate. Binding *et al.* (1), also starting with mesophyll protoplasts, found that a MS medium supplemented with 0.55 mg/liter kinetin and 0.9 mg/liter IAA promoted shoot formation from protoplast-derived callus.

The maintenance and long-term storage of germplasm is an important aspect of any breeding program. Espinoza *et al.* (3) reported that potato internode shoots can be stored for long periods when cultured on a MS medium supplemented with 4% mannitol, 3% sucrose, and 0.8 g/liter agar. This media exerts an osmotic stress that reduces the growth rate and induces proliferation of shoot internodes, thus providing many nodes for mass propagation when the material is needed. Reducing the storage temperature to 8°C significantly reduced the growth rate of the plantlets; thus transfers under these conditions are only required every 2–3 years.

Materials Required
1.   6 small-mouth Mason jars
2.   1000-ml beaker and 250-ml beaker
3.   20 sterile plastic or glass petri plates
4.   Bunsen or alcohol burner
5.   3 pairs of forceps and 3 scalpels
6.   Waterproof marking pen and labels
7.   Culture tubes (20 × 150 mm) with closures and slant racks to hold them
8.   125-ml Erlenmeyer flasks
9.   1000 ml of 20% Clorox solution supplemented with a few drops of Tween-20
10.   1000 ml of sterile distilled water
11.   200 ml of 95% ethanol
12.   Solid shoot multiplication medium containing:

MS salts (Macro- and micronutrients)[1]		
Sucrose	30.0	g/liter
Agar	8.0	g/liter

[1] Prepared as described in Chapter 2 or prepackaged salts.

GA$_3$ 0.25 mg/liter
Calcium pantothenic acid 2.0 mg/liter
MS Modified Vitamin Mixture (sigma 1.0 ml/liter
cell culture)

13. Liquid shoot multiplication medium containing:

MS salts (Macro- and micronutrients)
Sucrose 20.0 g/liter
GA$_3$ 0.4 mg/liter
BA 0.5 mg/liter
NAA 0.01 mg/liter
Calcium pantothenic acid 2.0 mg/liter
MS Modified Vitamin Mixture 1.0 ml/liter

14. Germplasm storage medium containing:

MS salts (Macro- and micronutrients)
Sucrose 30.0 g/liter
Mannitol 40.0 g/liter
Agar 8.0 g/liter
MS Modified Vitamin Mixture 1.0 ml/liter

15. 6 healthy medium-sized Irish potatoes

Procedures
1. Place the basal end of a medium-sized Irish potato in a Mason jar filled with water; repeat with six potatoes. Allow several weeks for the sprouting of eyes to occur.
2. Prepare the liquid and solid multiplication medium as described above. Dispense 10 ml of the solid medium into each culture tube and 25 ml of the liquid medium into each 125-ml Erlenmeyer flask and then sterilize. Prepare 10 replicates of each medium.
3. Remove sprouts, sterilize them for 10 min in 20% Clorox solution, and then rinse the tissue three times with sterile distilled water.
4. Aseptically section the sprouts into 10-mm sections with each section containing one node. Inoculate one node per culture tube or four nodes per culture flask. Incubate cultures under low-light conditions at 25°C. Liquid cultures should be agitated at 80–100 rpm.
5. Three weeks after inoculation, remove shoots from the culture vessels, divide into sections containing one node, and reinoculate onto fresh medium as described in the previous step. Repeat this procedure every 3 weeks (Fig. 23.1).

Germplasm Storage

1. Prepare the germplasm storage medium as described above. Dispense 10 ml of the medium into each of 20 culture tubes and sterilize.
2. Collect shoots from the micropropagation stage at least 2 weeks after the previous subculture. Divide the shoots into sections containing one node and place one section into each tube containing germplasm storage medium.
3. Store cultures in low light at 8°C.

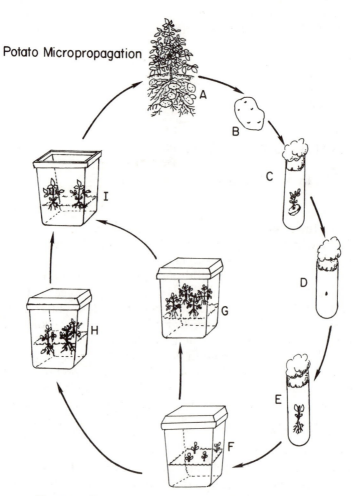

Fig. 23.1.   Various stages of potato micropropagation.

Removal from Storage and Rooting of Shoots

1. Randomly select cultures from germplasm storage every 4 months. Collect single nodes from the *in vitro*-derived plantlets and subculture twice on solid shoot multiplication medium in culture tubes.
2. When the plantlets are 3–5 cm high and have developed a good root system, transplant them into peat pots containing a suitable soil mixture (3 bark : 1 peat : 1 vermiculite). Take care not to damage roots.
3. Place peat pots with plantlets in an environment with a relatively high humidity for the first few days. Evaluate plantlets for genetic variation after 6 weeks (see Fig. 23.1).

Scheduling

Event	Timing
Initiation of cultures on multiplication medium	Day 0
Subculture of plantlets on multiplication medium	Day 21, 42, 63, etc.
Transfer to germplasm storage medium	2 weeks after previous subculture
Removal from storage and reculture on multiplication medium	As required—at least once every 4 months
Transfer of plantlets to soil mixture	When well-informed root system is present on shoot—ca. 4 weeks

Recording Results

1. Record all details of setting up the experiment.
2. Calculate the number of shoots produced from tissue growing on both a solid and liquid medium.
3. Determine the amount of variation (somaclonal variation) in the shoot tissue stored at 8°C.

REFERENCES

1. Binding, H., R. Nehls, O. Schieder, S. Sopory, and G. Wenzel. 1978. Regeneration of mesophyll protoplasts isolated from dihaploid clones of *Solanum tuberosum*. Physiol. Plant. 43:52.
2. Dunwell, J. M. and N. Sunderland. 1973. Anther culture of *Solanum tuberosum* L. Euphytica 22:317.
3. Espinoza, N., R. Estrada, P. Tovar, J. Bryan, and J. H. Dodds. 1984. Tissue Culture Micropropagation, Conservation and Export of Potato Germplasm. Specialized Technology Document 1. International Potato Center, Lima, Peru.

4. Foroughi-Wehr, B., H. M. Wilson, G. Mix, and H. Saul. 1977. Monohaploid plants from anthers of a dihaploid genotype of *Solanum tuberosum* L. Euphytica 26:361.

5. Lam, S. L. 1975. Shoot formation in potato tuber discs in tissue culture. Am. Potato J. 52:103.

6. Mellor, F. C. and R. Stace-Smith. 1969. Development of excised potato buds in nutrient cultures. Can. J. Bot. 47:1617.

7. Roest, S. and G. S. Bokelmann. 1976. Vegetative propagation of *Solanum tuberosum* L. Potato Res. 19:173.

8. Shepard, J. F. and R. E. Totten. 1977. Mesophyll cell protoplasts of potato: Isolation, proliferation and plant regeneration. Plant Physiol. 60:313.

9. Skirvin, R. M., S. Lam, and J. Janick. 1975. Plantlet formation from potato callus *in vitro*. HortScience 10:413.

PART IV

Protoplast Culture

24

Overview of Protoplast Isolation and Culture

One of the most significant events in the field of plant tissue culture during recent years has been development of techniques for the isolation, culture, and fusion of protoplasts, i.e., of cells devoid of their retaining walls. Protoplast culture techniques are especially important because of their potential application in efforts to improve plant species by cell modification and somatic hybridization.

Research to date has shown that protoplasts in culture can be induced to regenerate into entire plants, to undergo intra- and interspecific fusions to form somatic hybrids, and to take up foreign organelles and genetic materials. Although the technology of protoplast culture is still very much in the development stage, protoplasts already have been used to investigate problems in plant physiology, virology, pathology, and cytogenetics. In the future, protoplast culture is likely to be one of the most frequently used research tools for tissue culture studies and may have unlimited potential in genetic engineering for plant improvement.

It is important to stress that the procedures used to isolate protoplasts can significantly affect the subsequent behavior of the protoplasts. Protoplasts must be kept in a suitable plasmoyticum to prevent rupture of the membranes.

ISOLATION OF PROTOPLASTS

Protoplasts are isolated primarily by mechanical or enzymatic methods. Mechanical isolation is used only occasionally today, but this technique remains historically important. With enzymatic methods of isolation, large quantities of protoplasts can be obtained, little cell breakage results, and much less osmotic shrinkage occurs than with mechanical methods.

From Leaves

The preparation of protoplasts from leaves involves four basic steps:
(1) sterilization of leaves, (2) removal of epidermal cell layer, (3) enzymatic treatment in an osmoticum, and (4) isolation of protoplasts by filtration and centrifugation (Fig. 24.1).

Fully expanded leaves from young vegetative tissue are sterilized by dipping in 70% ethanol for 1–5 min. They are then surface-sterilized in a 1–2% solution of sodium hypochlorite for 10–30 min. Subsequent operations are carried out under aseptic conditions.

The lower epidermis is removed by carefully peeling back the epidermal region or by scraping off the epidermal region if it cannot be peeled

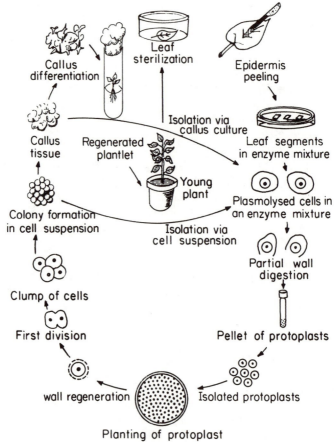

Fig. 24.1. Steps involved in the isolation, culture, and regeneration of protoplasts from leaves.

back. The leaves are generally easy to peel if the water supply to the plant is limited before excision of the leaves or if the leaves are allowed to become flaccid after sterilization. The peeled leaves are cut into small sections before enzyme treatment.

An osmoticum is used to stabilize and prevent rupture of the protoplast membranes. The compounds commonly used as osmoticums are sucrose, glucose, sorbitol, or mannitol, at concentrations of 8–20% or 0.3–0.5 M. Inorganic salts (e.g., KNO_3, KH_2PO_3, $MgSO_4$, and $CaCl_2$) also may be used to stabilize the membranes. Concentrations of 0.1–1.0 mM of $CaCl_2$ and $MgSO_4$ commonly are used.

The enzyme solution, prepared in an appropriate osmoticum, contains one or more of the following: pectinase (0.1–1.0%), cellulase (1.0–10%), and drieslase or hemicellulase (0.1–1.0%). The pH of the enzyme solution should be adjusted to 5.4–5.7 before use.

One-Step Isolation Method

In the direct (one-step) method, peeled segments of leaves are placed with the lower surface downward in a petri dish containing about 10 ml of filter-sterilized enzyme solution (mixture of pectinase, cellulase, and drieslase or hemicellulase). The leaf segments are incubated at 25°C for 15–18 hr. After incubation, the leaf segments can be gently teased with a pipette to liberate the protoplasts. Next, the protoplasts are filtered through a 63–69 μm mesh screen to remove leaf debris. The filtrate containing the protoplasts is centrifuged at 100g for 1 min. The protoplasts form a pellet; the supernatant and cell debris are decanted off. This washing process is repeated three times. The wash solution is generally made up of an osmoticum and may also contain a salt component such as MS stock salts used at 1/10 or 1/2 strength. During the final rinse, the osmoticum should be replaced with a 20% sucrose solution. After the final centrifugation at 200g for 1 min, the protoplasts will float and should be pipetted off.

Two-Step Isolation Method

In the alternative sequential (two-step) method, peeled leaf pieces are placed in a 0.5–2% pectinase solution containing an appropriate osmoticum (pH 5.8), vacuum-infiltrated in a desiccator for 5 min, and then incubated in a water bath at 25°C with gentle shaking for 15 min; after this, the mixture is incubated for another 60 min. At this stage, a drop of cells should be examined under a microscope to determine if the cells have separated completely. If not, continue incubation for another 30–60 min. After separation is achieved, the mixture is filtered to remove large cellular debris; the filtrate is centrifuged and washed three times with an appropriate osmotic solution as described in the preceding section.

The cell pellet obtained after washing is dispersed into a second enzyme solution (1–4% cellulase + 13% osmoticum, pH 5.4) and incubated for about 90 min at 30°C. After incubation, the cells are centrifuged at 100g for 1 min. The protoplast pellet is then washed three times with the osmoticum. In the final step, the protoplasts are isolated and purified by floating them in a 20% sucrose solution, as described in the one-step method.

In the one-step method, both spongy mesophyll and palisade mesophyll protoplasts are isolated, whereas in the two-step method, only palisade mesophyll protoplasts are obtained. Because of this, the yield of protoplasts is greater with the one-step method than with the two-step method. However, the protoplasts isolated with the two-step method are generally of a better quality and have a higher viability. It appears that the long incubation period used in the one-step method (15–18 hr) adversely alters the plasma membrane compared to the two-step method, which has an incubation period of approximately 3 hr.

From Callus Cultures

Actively growing young callus cultures are ideal material for the isolation of large quantities of protoplasts. The isolation procedure is basically the same as that used with leaves. However, the optimal concentration of enzymes, particularly of cellulase, may be less with callus than with leaf tissue. Also, the time required for incubation is generally only 4–16 hr with callus tissue. Cultures should be incubated at 30°–33°C during isolation. Older callus cultures tend to form giant cells with thick cell walls, which may be difficult to digest. Therefore, young actively growing callus cultures should be used, ones that have been subcultured no more than 2 weeks before protoplast isolation.

From Cell Suspension Cultures

Cell suspension cultures that are young and actively growing also provide an excellent source material for isolating protoplasts. In a typical isolation procedure, 5 ml of a high-density cell suspension is transferred to a 1-ml conical centrifuge tube and centrifuged at 100g for 1–2 min. After the supernatant is removed, the cells are resuspended in wash medium and centrifuged again. The supernatant is removed then cells are resuspended in 5 ml of enzyme (14% cellulase + 0.5–2% pectinase) and poured into a petri plate. The plate is sealed with Parafilm, placed on a platform shaker at 30–75 rpm, and incubated for 2–6 hrs.

It may be desirable to add a very low concentration of cellulase and pectinase (0.1 and 0.01%, respectively) to the cell suspension to discourage the formation of aggregates and thick-walled cells. Consistently better yields are obtained when efforts are made to prevent the formation of aggregates.

PROTOPLAST CULTURE AND REGENERATION

The basic phenomenon that plant cells are totipotent and have the capacity to regenerate whole plants makes protoplast culture possible. The first step in protoplast culture involves the regeneration of the cell wall around the protoplast membrane. Once the cell wall has formed, cell division must be induced in the new cell. Slowly a small colony of cells will form. By manipulation of the nutritional and physiological conditions, the experimenter may induce this cultured tissue to continue callus growth or regenerate an entire plant (Fig. 24.1). Nagata and Takebe (5) described the regeneration of whole plants from mesophyll protoplasts of tobacco isolated with a two-step enzyme method. Spontaneously fused tobacco leaf protoplasts have undergone cell wall regeneration and cell division when the medium is supplemented with 23% sucrose instead of mannitol.

General Culture Methods

The culturing of protoplasts presents many special problems. Thus, it is important that an efficient, reproducible culture method be developed for each species. Protoplasts may be cultured either in a liquid or solid medium.

Methods Using a Liquid Medium

Several common methods for culturing protoplasts in a liquid medium are summarized below:

1. Protoplasts may be incubated in a thin layer of nutrient medium in a petri plate. The dish should be sealed with Parafilm and incubated under low-light or dark conditions at 25°–28°C.
2. Protoplasts may be cultured in small drops of medium placed in a petri plate. The plate is sealed, placed in a humidity box, and incubated at 25°–28°C under low-light intensities (1).
3. Protoplasts may be incubated in a 50- or 100-ml Erlenmeyer flask containing 5 ml of culture medium. The flasks are incubated statically at 25°C at a light intensity of 2500 lux.
4. Protoplasts may be cultured in microchamber slides or microwell plates. The protoplasts are incubated at 25°–28°C under low-light and high-humidity conditions.

Methods Using a Solid Medium

Karatha *et al.* (*3*) and Gamborg *et al.* (*2*) have described procedures for suspending isolated protoplasts in a liquid medium and then plating them on the top of a solid agar medium for further development (Fig. 24.1). Perhaps the best method for culturing protoplasts is to plate them on a solidified agar medium, using the technique developed by Nagata and Takbe (*5*) for plating out tobacco cells. With this method, a large number of protoplasts or cells can be conveniently handled and observed under a microscope, and plating efficiency is easily determined. In this method, 2-ml aliquots of isolated protoplasts (density of ca. 10^5 cells/ml) are centrifuged and the protoplasts resuspended in a liquid medium. The protoplast suspension is poured into a small petri plate with an equal volume of a nutrient medium containing 1–2% agar. At the time of mixing, particular care should be taken to ensure that the temperature of the medium does not exceed 45°C. The petri plates are sealed with Parafilm to prevent desiccation and the agar allowed to solidify. Culture plates are incubated in an inverted position at 25°–28°C.

Components of Protoplast Culture Media

The components required in protoplast culture media are generally the same as those used in media for callus and cell suspension cultures. The one major difference is the inclusion of an osmoticum, in both the protoplast isolation and culture media, to replace the wall pressure exerted by the cell wall in the intact cell. Without the presence of an osmoticum in the medium, protoplast membranes would quickly rupture. It is well documented that medium leaf osmotic pressures may be significantly influenced by environmental conditions (*6*). One approach to reducing medium osmotic pressures is to subject the plant material to lengthy dark periods before isolation. Reduced osmotic potentials are usually achieved by the addition of mannitol, sorbitol, glucose, or sucrose to the enzyme mixture.

Research has shown that the iron, zinc, and ammonium concentrations commonly used in tissue culture media may be too high for protoplast culture media. Pea mesophyll protoplasts seem to have an optimum iron and zinc requirement of 50 μM and 10 μM, respectively. Increasing the calcium concentration in a protoplast culture medium two to four times over the normal concentrations of a cell culture medium may be beneficial in preserving membrane integrity.

Uchimiya and Murashige (*7*) showed that tobacco protoplast cultures required substantially less sugar than did cell cultures. The optimum sugar level was 1.5% for both reformation of the cell wall and the resumption of cell division. The protoplasts grew equally well on sucrose and glucose. Most of the media developed for protoplast culture contain 3–5% sugar, which is about the same as that used in tissue culture media.

The vitamin requirements for protoplast culture media are usually the same as those for standard tissue culture media. Thiamin, myo-inositol, nicotinic acid, and pyridoxine are usually included.

Organic nitrogen in the form of casein hydrolysate is generally included in protoplast culture media to supply a form of reduced nitrogen immediately available to the cells.

Auxins and cytokinins are generally required to induce cell wall formation and cell division in protoplast cultures. Significant increases in the number of protoplasts that form cell walls are generally noted when both an auxin and cytokinin are included in the protoplast culture medium. Most protoplast culture media will include a combination of both auxins and cytokinins instead of relying on a single auxin or cytokin source to induce cell wall formation and division. Auxin sources commonly used in protoplast culture media include 2,4-D, NAA, and IAA. While 2,4-D may be beneficial in stimulating cell wall formation and cell division alone, its use as the sole auxin source may lead to the loss of morphogenetic potential by the developing microcalli. In most media at least two of the cytokinins, kinetin, BA, 2iP, or zeatin, will be included. The use of zeatin seems to enhance cell division and morphogenetic potential of the developing microcalli.

Cell Wall Formation

Protoplasts in culture generally start to regenerate a cell wall within a few hours after isolation and may take two to several days to complete the process under suitable conditions. The newly synthesized cell wall can be demonstrated by staining it with 0.1% Calcofluor white fluorescent stain (Sigma Chemical).

When protoplasts are cultured on a solid medium by the plating out technique described earlier, both protoplast density and light intensity influence plating efficiency. A protoplast density of $0.5–1.0 \times 10^5$ cells/ml is required to achieve high plating efficiencies. The plating efficiency of tobacco protoplasts has been shown to be considerably enhanced if they are incubated for the first two days under low light conditions (300 lux) and then transferred to a high-light regime (3000 lux).

Growth and Division

Regeneration of a cell wall is not necessarily a prerequisite for the initiation of nuclear division in protoplast cultures. Cell wall formation is, however, required before cytokinesis occurs. Once a cell wall is formed, the reconstituted cells show a considerable increase in cell size and the first cell division generally occurs within 3–5 days. The second division will occur within a week; by the end of the second week in culture, small aggregates of cells will be present. After 3 weeks, small cell colonies will be visible and colonies of *ca.* 1 mm are present after approximately 6 weeks in culture. Once small colonies have formed, their further growth is inhibited if they are allowed to remain on the original high osmotic medium. The colonies should, therefore, be transferred to a mannitol or sorbitol-free medium.

PROTOPLAST FUSION AND SOMATIC HYBRIDIZATION

One of the most important practical uses of protoplasts is for somatic hybridization, a process that is especially important in sexually incompatible plants and in cases where conventional methods of breeding fail to operate. Protoplasts can fuse spontaneously during isolation or fusion can be induced by exposure to special conditions.

During isolation, spontaneous fusion can occur between two or more adjacent protoplasts. The process appears to take place when the plasmodesmata between adjoining protoplasts expands rather than breaks, with the result that the nuclear and cytoplasmic material of the protoplasts fuse into one unit. The phenomenon occurs at a greater frequency with protoplasts isolated from cell cultures than with protoplasts isolated from leaves. Following fusion, the resulting polynucleate cell (polykaryon) may reform a cell wall and the nuclei may enter mitosis synchronously.

Fusion of protoplasts from different cell sources seldom occurs spontaneously and must be induced by some treatment that will establish a close contact between the protoplasts, yet not damage the integrity of the cell. If the treated protoplasts have been isolated from different plants, some of the resulting fused cells will contain cellular material originating from the different sources, that is, they will be hybrids. The fused, hybrid protoplasts can be plated and hybrid colonies selected and transferred to callus culture media. Under suitable conditions, plant regeneration may occur to yield somatic hybrid plants. This process of somatic hybridization resulting from protoplast fusion is illustrated in Fig. 24.2.

Methods of Inducing Protoplast Fusion

Several of the methods for inducing protoplast fusion are described briefly in this section.

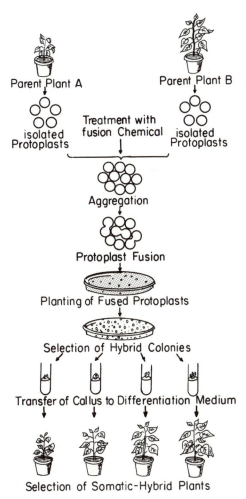

Parent Plant A

Parent Plant B

Treatment with
fusion Chemical

isolated
Protoplasts

isolated
Protoplasts

Aggregation

Protoplast Fusion

Planting of Fused Protoplasts

Selection of Hybrid Colonies

Transfer of Callus to Differentiation Medium

Selection of Somatic-Hybrid Plants

Fig. 24.2. Steps involved in protoplast fusion and somatic hybridization to produce hybrid plants.

Mechanical Treatment

Isolated protoplasts to be fused are brought together or in intimate contact through the use of micromanipulators and perfusion micropipettes. Part of the micropipette tip is blocked so that the protoplasts are retained and compressed by the flow of the liquid. The number of cells induced to fuse by this method is very low, and thus this method is not commonly utilized.

Treatment with Sodium Nitrate

Isolated protoplasts are suspended in a mixture of 5.5% sodium nitrate in a 10% sucrose solution. The solution containing the protoplasts is incubated in a water bath at 35°C for 5 min and then centrifuged for 5 min at 200g. Following centrifugation, most of the supernatant is decanted and the protoplast pellet is transferred to a water bath at 30°C for 30 min. During this period, most of the protoplasts undergo cell fusion. The remaining aggregation mixture is decanted gently and replaced with the culture medium to be used supplemented with 0.1% NaNO$_3$. The protoplasts are left undisturbed for some time after which they are washed twice with the culture medium and plated.

Calcium Ions at High pH

Keller and Melchers (4) evaluated the effect of high pH and calcium ions on the fusion of tobacco protoplasts. In their method, isolated protoplasts are centrifuged for 3 min at 50g in a fusion-inducing solution of 0.5 M mannitol containing 0.05 M CaCl$_2 \cdot$ 2H$_2$O at a pH of 10.5. The centrifuge tubes containing the protoplasts are then incubated in a water bath at 37°C for 40–50 min. After this treatment, 20–50% of the protoplasts were involved in fusion.

Polyethylene Glycol (PEG)

One of the most successful techniques for fusing protoplasts is to suspend the cells in a solution containing PEG, which enhances the agglutination and fusion of protoplasts in several species. When sufficient quantities of protoplasts are available, 1 ml of the protoplasts, suspended in a culture medium, are mixed with 1 ml of a 56% PEG solution. The tube is then shaken for 5 sec and allowed to settle for 10 min. The protoplasts are then washed several times to remove the PEG and then suspended in their culture medium.

When only microquantities of protoplasts are available, the protoplasts must be fused using the drop culture technique. Protoplasts of two different sources are mixed in equal quantities and 4–6 microdrops (100 μl each) are placed in small petri dishes and allowed to settle for 5–10 min at room temperature. Then 2–3 microdrops (50 μl each) of PEG are added to the periphery of the mixed protoplast solution to induce fusion. Placing the microdrops containing the protoplasts onto a coverslip located within the petri dish may aid in the transfer of the cells at a later stage. This procedure also prevents the protoplasts from sticking to the surface of the petri dish, and allows the researcher to fix, stain, and observe the protoplasts while they are on the coverslip.

REFERENCES

1. Constable, F. 1975. Isolation and culture of plant protoplasts. pp. 11–21. *In* Plant Tissue Culture Methods. O. L. Gamborg and L. R. Wetter (editors). Nat. Res. Council of Canada, Saskutoon.

2. Gamborg, O. L., J. P. Shyluk, and K. K. Kartha. 1975. Factors affecting the isolation and callus formation in protoplasts from shoot apices of *Pisum sativum* L. Plt. Sci. Lett. 4:285–292.

3. Kartha, K. K., M. R. Michayluk, K. N. Kao, O. L. Gamborg, and F. Constabel. 1974. Callus formation and plant regeneration from mesophyll protoplasts of ripe plants (*Brassica mapus* cv. *Zephyr.*). Plt. Sci. Lett. 3:265–271.

4. Keller, W. A. and G. Melchers. 1973. The effect of high pH and calcium on tobacco leaf protoplast fusion. Z. Naturforsch. 288:737–741.

5. Nagata, T. and I. Takebe. 1970. Cell wall regeneration and cell division in isolated tobacco mesophyll protoplasts. Planta 92:301–308.

6. Shepard, J. F. and R. E. Totten. 1975. Isolation and regeneration of tobacco mesophyll cell protoplasts under low osmotic conditions. Plant Physiol. 55:689–694.

7. Uchimiya, H. and T. Murashige. 1976. Influence of the nutrient medium on the recovery of dividing cells from tobacco protoplasts. Plant Physiol. 57:424–429.

25

Isolation and Culture of Protoplasts from Carrot Cell Suspension Cultures

Early studies indicated that carrot (*Daucus carota*) cells can be cultured in a liquid medium and their totipotency expressed through somatic embryo differentiation. The successful isolation of protoplasts from carrot cell suspension cultures and the redifferentiation of plantlets from protoplast-derived calli also has been reported (1). Later modifications in the hormone components of the culture medium resulted in induction and formation of somatic embryos without the formation of callus as an intermediate stage. Because of the high morphogenetic potential of carrot cell suspension cultures, carrot cell lines offer an excellent experimental system for the analysis of plant cell growth and differentiation and protoplast-mediated genetic manipulation of plants.

The efficiency of protoplasts will be highly dependent on the type and condition of the cell line. The use of newly established cell lines, preferably less than a year old, is suggested if plant regeneration is a basic requirement in the research. Morphogenetic potential is reported to be more stable in wild carrot cultures than in domesticated carrot cultures. Protoplasts isolated from actively dividing cell cultures generally show a higher frequency of protoplast division and growth than do protoplasts from older cultures in the stationary phase.

Materials Required
1. Culture tubes (20 × 150 mm) for callus cultures
2. 125-ml Erlenmeyer flasks for cell suspension cultures
3. Sterile glass or plastic petri plates (15 × 100 mm)
4. Sterile Pasteur pipettes
5. Forceps and scalpels
6. Sterile graduated centrifuge tubes
7. Bunsen or alcohol burner
8. Variable speed shaker

9. Nylon or wire mesh sieves (ca. 40–60 μM)
10. 0.1% mercuric chloride solution
11. Callus induction medium (1 liter)

B-5 basal salts (macro- and micronutrients)[1]	
Thiamin · HCl	1.0 mg/liter
Myo-inositol	100.0 mg/liter
Sucrose	30.0 g/liter
2,4-D	1.0 mg/liter
BA	0.1 mg/liter
Agar	8.0 g/liter

12. Cell suspension medium (1 liter)

B-5 basal salts (macro -and micronutrients)[1]	
Thiamin · HCl	1.0 mg/liter
Myo-inositol	100.0 mg/liter
Nicotinic acid	0.5 mg/liter
Pyridoxine · HCl	0.5 mg/liter
Sucrose	30.0 g/liter
2,4-D	1.0 mg/liter
BA	0.1 mg/liter

13. Protoplast enzyme/isolation medium containing the following components with final pH of 5.6:
    2% Onozuka R-10 Cellulast or 2% cellulysin
    1% Pectinase
    0.5% Driselase
    0.5% Rhozyme
    0.35 M Sorbitol
    0.35 M Mannitol
    3 mM 2-(N-morpholino) ethane-sulfonic acid (MES)
    6 mM CaCl$_2$ · 2H$_2$O
    0.7 mM NaH$_2$PO$_4$ · H$_2$O
14. Protoplasts culture medium

B-5 basal salts (macro- and micronutrients)[1]	
Thiamin · HCl	1.0 mg/liter
Myo-inositol	100.0 mg/liter
Nicotinic acid	0.5 mg/liter
Pyridoxine · HCl	0.5 mg/liter
Glucose	68.4 g/liter
2,4-D	0.5 mg/liter
Zeatin	0.11 mg/liter

[1] Prepared as described in Chapter 2 or prepackaged salts.

15. Protoplasts wash medium

B-5 basal salts (macro- and micronutrients)

Thiamin·HCl	1.0 mg/liter
Myo-inositol	100.0 mg/liter
Nicotinic acid	0.5 mg/liter
Pyridoxine·HCl	0.5 mg/liter
Sorbitol	0.35 M
Mannitol	0.35 M
$CaCl_2 \cdot 2H_2O$	6 mM

16. Ca. 100 healthy carrot seed

Callus Induction and Cell Suspension Culture
1. Surface-sterilize the seed by placing them in a mercuric chloride solution for 5–10 min and then washing three times with sterile distilled water. Inoculate the sterilized seed onto a basal medium consisting of MS salts and 8.0 g/liter agar.
2. Prepare callus induction medium as described in previous section. Dispense medium into culture tubes (10 ml/tube) and sterilize.
3. After the seed have germinated, remove the hypocotyls and cut into sections 1.0–1.5 cm long. Transfer one section to each callus culture tube.
4. Incubate the callus cultures at 25°C in the dark for 4–6 weeks. Callus can be maintained by subculturing the callus at intervals of 4–6 weeks.
5. Prepare liquid cell suspension medium as described in the previous section. Dispense this medium into 125-ml Erlenmeyer flasks (20–30 ml/flask) and sterilize.
6. Remove callus from 4- to 6-week old callus cultures and transfer 1–2 g of callus tissue to each flask containing liquid cell suspension medium.
7. Place culture flasks on a shaker set at 100–125 rpm and incubate at 25°C. Fresh medium should be added to replace the old culture medium every 7–10 days during the first period of culture. Later the cells should be subcultured twice a week. When the old culture medium is removed, large, nondividing cells should be discarded from the culture. The division frequency of the suspension culture and the subsequent protoplast culture is greatly enhanced if large, nondividing cells are removed.

Protoplast Isolation
1. Protoplasts should be isolated from suspension cultures that have been subcultured within the previous 2 to 3 days. First, concen-

trate the cell suspension by gently centrifuging the culture (800 rpm) and then removing some of the supernatant.

2. Add an equal volume of the protoplast enzyme/isolation solution to the concentrated cell suspension. Incubate the mixture on a gyratory shaker (50 rpm) at 25°C overnight.

3. After digestion, filter the cellular material through a nylon or wire sieve, collecting the filtrate in a centrifuge tube. Centrifuge at 800 rpm for 5–10 min.

4. Wash the protoplasts by removing the supernatant and resuspending the pellet in 10 ml of the protoplast wash medium described above. Repeat the centrifugation and wash process two additional times.

Protoplast Culture

1. Prepare protoplast culture medium and filter-sterilize it.

2. To the washed, isolated protoplasts, add sufficient medium to give a density of 1×10^5 to 5×10^6 protoplasts/ml. Dispense 5 ml of this suspension into a petri dish containing 10 ml of the protoplast culture medium at 35–40°C and seal with Parafilm.

3. Incubate protoplast cultures at 25°C in the dark. After 10–14 days, add a few drops of medium to the petri dishes to replenish the medium.

Embryo Differentiation and Plantlet Regeneration

1. To initiate embryo formation and plantlet regeneration follow steps 4–8 in the Procedures section of Experimental Exercise 18 (Chapter 21).

Scheduling

Event	Day
Initation of callus cultures	0
Subculture of callus	Every 60 days
Initiation of suspension cultures	1–2 weeks after the last callus subculture
Subculture of cell suspension cultures	Every 14 days
Initiation of protoplasts isolation	2–3 days after last cell suspension culture transfer
Plating of protoplast	Day of isolation
Visual appearance formation of protoplast-derived micro-calli	24 wks after protoplast isolation

Recording Results
1. Record all procedures of setting up each phase of the study.
2. Calculate cell suspension growth and select the most actively growing culture for protoplast isolation.
3. Determine the number of protoplasts released/gram of cell suspension.
4. Calculate the percentage of protoplasts which form micro-calli colonies.

REFERENCES

1. Dudits, D. 1984. Isolation and culture of protoplasts from carrot cell suspension cultures. *In* Cell Culture and Somatic Cell Genetics. I. K. Vasil (editor). Academic Press, New York.

26

Isolation and Culture of Protoplasts from Grass Callus Cultures

Until recently, there has been little interest in the *in vitro* propagation of turfgrasses because these species generally are easily propagated and would not be economically feasible to clone. However, protoplast manipulation and somatic hybridization may be useful in efforts to develop genetically superior turfgrass cultivars. For example, cultivars capable of tolerating adverse conditions such as heat and drought would be economically valuable for southern golf greens.

There are several ways in which *in vitro* techniques may be used for the genetic improvement of plants when conventional breeding methods fail due to sexual incompatibility or other reasons. In one approach, isolated protoplasts are stressed by various treatments that may induce useful genetic changes in the regenerated plantlets. Another approach is to induce fusion of protoplasts from two or more cultivars, creating a somatic hybrid that may have desirable characteristics (see Fig. 24.2). Both of these approaches depend on development of systems for the isolation, culture, and regeneration of protoplasts from the species of interest.

Protoplasts have been successfully isolated, cultured, and regenerated from a number of species (e.g., tobacco, carrot, petunia, beans), but the grass species have proved difficult to work with. The isolation of mesophyll protoplasts and regeneration of plantlets from them within the grass family was first achieved with *Bromus* (3) and sugarcane (6). Successful isolation and culture of protoplasts has more recently been reported in barley (4), rice (1, 2), *Tritecum monococcum* (7), pearl millet (9), napier grass (10), and corn (8). These reports provide hope that the degree of success achieved in isolating and culturing protoplasts from dicotyledonous plants may also be obtained with the monocotyledonous grasses.

Materials Required
1. 20 sterile glass or plastic petri dishes
2. 10 sterile and capped 50-ml Erlenmeyer flasks

3. Sterile (.45 μm) filtration unit
4. Top-loading electronic balance
5. Variable speed shaker
6. Wire meshes (approximately 100 μm and 65 μm mesh size)
7. Sterile Pasteur pipettes
8. 2 multiwell dishes with 96 or 24 wells
9. 10 sterile conical graduated centrifuge tubes
10. Culture tubes (20 × 150 mm)
11. Parafilm
12. Alcohol or bunsen burner
13. Forceps, spatulas, and scalpels
14. 4 250-ml beakers
15. 200 ml 95% ethyl alcohol
16. 1000 ml of 20% Clorox solution
17. 1000 ml of sterile distilled water
18. Callus induction medium

MS basal salts (macro- and micro nutrients)[1]

Agar	8.0 g/liter
Sucrose	30.0 g/liter
Thiamin · HCl	1.0 mg/liter
Myo-inositol	100.0 mg/liter
2,4-D	5.0 mg/liter
BA	1.0 mg/liter

19. Protoplasts isolation medium

MS basal salts (macro- and micronutrients)[1]

Thiamin · HCl	1.0 mg/liter
Myo-inositol	100.0 mg/liter
Mannitol	10%
Cellulase Type II (Sigma Chemical)	7%
Pectinase (Sigma Chemical)	2%
HEPES	3 mM

20. Protoplast wash medium consisting of half-strength MS salts and 10% mannitol (pH 5.8)
21. Protoplast culture medium

MS salts[1]

Thiamin · HCl	1.0 mg/liter
Myo-inositol	100.0 mg/liter
2,4-D	5.0 mg/liter
BA	1.0 mg/liter
Zeatin	1.0 mg/liter

[1] Prepared as in Chapter 2 or prepackaged salts.

Mannitol	1.5%
Sucrose	1.5%
Agar	8.0 g/liter

22. Caryopses of *Agrostis tenuis* Huds 'Highland' and *Agrostis palus-tris* Huds 'Seaside'

Callus Induction

1. Prepare callus induction medium as described in the previous section. Dispense into culture tubes (10 ml/tube) and sterilize.
2. Secure a 1-g sample of caryopses from each of the grass species. Disinfect and inoculate each of the samples as described in steps 3 and 4 of the Procedures section of Chapter 13.
3. Culture for 4–8 weeks and subculture if necessary. Do not use callus older than 12 weeks for protoplast isolation.

Protoplast Isolation

1. Prepare the isolation medium by adding the dry ingredients (mannitol, cellulase, HEPES, and pectinase) to a 50-ml volumetric flask. Add about 3.5 ml of a liquid MS medium and swirl. While mixing, adjust the pH of the medium to 3.0; then readjust the pH to 8.5; finally bring the pH to 5.8. (This procedure enhances the mixing and dissolving of the enzymes.) Bring to final volume with liquid MS medium water. Filter-sterilize the medium and store it in sterile flasks.
2. Prepare protoplast wash medium as described in Materials section.
3. Aseptically weigh out 1.0 g of callus tissue and transfer to a sterile petri dish. Add 5.0 ml of protoplast isolation medium and macerate the callus using a sterile scalpel. Add another 5.0 ml of isolation medium, seal the petri dish with Parafilm, and incubate on a shaker (50–75 rpm) at 25°C for 12 hr.
4. After incubation, filter the protoplast suspension in the petri dish, first through a 100-μm wire mesh and then through a 65-μm mesh, to remove cell debris and large cell aggregates.
5. Determine the total volume of the filtrate containing the protoplasts. Remove a 0.5-ml aliquot and determine the protoplast density (protoplasts/ml) with a haemocytometer (Appendix 9) or cell-counting chamber.
6. Centrifuge the protoplast suspension at 500g for 5 min. Decant the supernatant and resuspend the protoplast pellet in wash medium. Repeat this procedure at least twice to wash the protoplasts.
7. After the final wash and centrifugation, resuspend the protoplast pellet in enough wash medium to give a density of $1–5 \times 10^5$ cells/ml. The actual volume of medium needed will depend on

the number of protoplasts in the pellet (determined from the cell count in step 5) and the final density at which you wish to plate the protoplasts.

Protoplast Culture
1. Prepare protoplast culture medium as described in Materials section. Maintain medium at 40°C during the plating procedure.
2. Place 1–2 ml of the culture medium and 1–2 ml of the protoplast suspension (adjusted to desired density) in a sterile petri dish. The final plating density should be $0.5–1.0 \times 10^5$ cells/ml.
3. Allow the agar medium to solidify, then seal the plates with Parafilm and incubate them in an inverted position at 25°C in the dark.
4. Alternatively, the protoplasts can be cultured in a liquid medium in a multiwell plate. If you use this procedure, prepare the protoplast culture medium as described but omit the agar. Place equal volumes (a few drops each) of culture medium and protoplast suspension in each well of the plate. Incubate at 25°C in the dark.
5. Inspect the cultures periodically under an inverted microscope. Colony formation should be evident within 30 days.

Scheduling

Event	Day
Initiation of callus from caryopses	0
Subculture of callus	Every 60 days
Isolation of protoplasts	1 week after last callus subculture
Culture of protoplasts	12–18 hours after
Appearance of protoplast-derived micro-calli	3–4 weeks after protoplast isolation

Recording Results
1. List all details of establishing the experiment.
2. Determine the number of protoplast released after enzymatic treatment.
3. Determine the percentage of protoplasts forming micro-calli.

REFERENCES

1. Cai, Q., Y. Quain, Y. Zhoa, and S. Wu. 1978. A further study on the isolation and culture of rice (*Oryza sativa* L.) protoplasts. Acta Bot. Sinica 20:97–102.
2. Deka, P. C. and S. K. Sen. 1976. Differentiation in calli originated from isolated protoplasts of rice (*Oryza sativa* L.) through plating technique. Theoret. Appl. Genet. 145:239–243.

3. Kao, K. N., O. L. Gamborg, M. R. Michayluk, and W. A. Weller. 1973. Colloq. Int. C.N.R.S. 212:207–213.

4. Koblitz, H. 1976. Isolation and cultivation of protoplasts from callus cultures of barley. Biochem. Physiol. Pfl. 170:287–293.

5. Lu, C. Y., V. Vasil, and I. K. Vasil. 1981. Isolation and culture of protoplasts of *Panicum maximum* Jaoq. (Guinea Grass): Somatic embryogenesis and plantlet formation., Z. Pflanzenphysiol. 104:311–318.

6. Maretzki, A. and L. G. Nickell. 1973. Formation of protoplasts from sugarcane cell suspensions and the regeneration of cell cultures from protoplasts. Colloq. Int. C.N.R.S. 212:51–63.

7. Nemet, G. and D. Dudits. 1977. Potential of protoplasts, cell and tissue culture in cereal research. pp. 145–163. *In* Use of Tissue Culture in Plant Breeding. F. J. Novak (editor). Czechoslovak Academy of Sciences, Institute of Experimental Botany, Prague.

8. Potrykus, I., C. T. Harms, H. Lorz, and E. Thomas. 1977. Callus formation from stem protoplasts of corn (*Zea mays* L.). Molec. Gen. Genet. 156:347–350.

9. Vasil, V. and I. K. Vasil. 1979. Isolation and culture of cereal protoplasts. I. Callus formation from pearl millet (*Pennisetum americanum*) protoplasts. Z. Pflanzenphysiol. 92:379–383.

10. Vasil V., D. Y. Wang, and I. K. Vasil. 1983. Plant regeneration from napier grass (*Pennesetum purpureum* Schun). Z. Pflanzenphysiol. 111:233–239.

PART V

Preparation of Specimens for Microscopy

27

Specimen Preparation
for Light Microsopy

In *vitro* cell cultures may form apical meristems (growing centers), vascular and/or tracheidal elements, or embryos spontaneously or in the presence of various induction stimuli. Microscopic examination of *in vitro* cultured tissues at various stages can help in determining the sequence of plantlet development.

The preparation of biological specimens for light microscopy can be divided into 5 major steps. These steps include fixation, dehydration, embedding, sectioning of the tissue, and affixing to the slide and staining. Fixation is probably the most critical step in the preparation of biological tissue (1). A properly fixed specimen should be as representative as possible of the living state. The primary method by which biological tissue is fixed for light microscopy. Coagulative fixatives including formalin-acetic acid-alcohol (FAA) and chromic acid-acetic acid-formalin (CRAF or Navaschin) are the primary fixatives used for light microscopy. Fixation of a specimen generally required submersion for at least 24 hours in the fixative. Most biological tissue can be stored indefinitely in the fixative.

The second major step in specimen preparation is dehydration. Most tissue must be dehydrated to remove water from the specimen before the tissue is infiltrated with the embedding medium. This dehydration process consists of treating tissue with a series of solutions containing progressively increasing concentrations of the dehydration agent and decreasing concentrations of water. Two methods are used to dehydrate the tissue. The first method dehydrates the tissue in a nonsolvent of paraffin (i.e., ethyl alcohol). Once the tissue is dehydrated, it is then transferred to a solvent of paraffin and embedded. The second method dehydrates the tissue in a solvent of paraffin (N-butyl alcohol or tertiary butyl alcohol) then embedded once dehydrated. The first method is commonly used where a plastic monomer is used as the embedding medium (1).

The paraffin or plastic matrix which tissues are embedded serves to support the tissue against the impact of the microtome knife as well as hold the tissue together after sectioning. A properly embedded tissue will have all cavities within the tissue as well as internal and external embedded surfaces with the matrix (1).

The staining of cellular structures is based on a specific affinity between certain dyes and particular cell structures. This specificity may be enhanced in certain cases with a mordant (usually a salt) which creates a three-way matrix between the dye, some part of the cell structure and the mordant (1). Excellent reviews on staining have been prepared by Conn (3) and Clark (2).

The procedures described in this chapter for the preparation of specimens for light microscopy have been used successfully with a variety of material. Alternative methods for fixation, dehydration, and staining are presented so the various methods may be evaluated.

Materials Required
1. 20 sterile plastic petri dishes
2. 2 250-ml beakers filled with 200 ml distilled water
3. Paraffin
4. Rotary microtome
5. Glass microscope slides (25 × 75 mm)
6. #2 glass coverslips (22 × 22 mm)
7. Several needles or probes
8. Hot plate (20°–80°C)
9. Slide trays
10. Wide-mouthed vials (20–30 ml) with caps
11. Pasteur pipettes
12. Slide containers with baskets for staining
13. 40 culture tubes (25 × 150 mm) each filled with 25 ml of MS medium supplemented with 1.0 mg/liter thiamin, 100 mg/liter myo-inositol, 0.1 mg/liter NAA, 0.1 mg/liter BA, 30.0 g/liter sucrose, and 8.0 g/liter agar
14. FAA fixative consisting of 50 ml formaldehyde, 50 ml glacial acetic acid, and 900 ml ethyl alcohol
15. Navashin's modified fixative consisting of the following solutions (keep separate):

    Solution A—1 g chromic acid, 7 ml glacial acetic acid, and 92 ml water
    Solution B—30 ml formaldehyde and 70 ml water

16. Absolute ethyl alcohol
17. Tert-butyl alcohol

18. Xylene
19. Chlorozol Black E, hematoxylin, aniline blue, safranin, and fast green dye powders (instructions for preparing dye solutions are given later)
20. 100 ml albumin (egg) fixative
21. 100 ml Permount
22. ca. 50 *Pisum* seed
23. 2 medium-size onions (obtain from curb market if possible so they will not be treated with anything)
24. 6 African violet leaves
25. 2 culture tubes containing actively growing callus

Preparation of Cultures and Source Materials
1. Sterilize and section African violet leaves as described in Chapter 6. Inoculate the 40 tubes containing supplemented MS medium with one leaf section/tube. Incubate tubes at 25°C under low-light conditions. Every 4 days until all 40 tubes have been used, select 4 tubes and remove small sections of the developing sprouts to prepare for histological examination as described in later sections.
2. Place 3 or 4 toothpicks in the sides of two fresh onions, one-fourth to one-half of the way up from the basal plate. Set each onion, basal part down, in a 250-ml beaker filled with water. The toothpicks should allow for air to reach the basal portion of each onion. Place the beakers in the dark and allow several days for rooting. Remove small sections of the sprouting roots for histological examination.
3. Sterilize about 50 *Pisum* seed using the procedure described in Chapter 16. Place 2 sheets of sterile filter paper in each of 20 sterile petri plates. Add several milliliters of sterile distilled water until the filter paper is thoroughly moistened. Place 4 of the sterile *Pisum* seed into each plate; seal each plate with Parafilm. Incubate the plates under dark conditions at 25°C for several days and then remove the distal 2–4 cm of the root tips and prepare for histological observation as described below.
4. Divide the callus tissue from two actively growing callus cultures into pealike segments about 1 cm in diameter.

Tissue Fixation
1. Prepare FAA fixative and Navashin's fixative (solutions A and B) as described in Materials section. Mix equal volumes of the Navashin's A and B solutions just before immersing tissue specimens into the fixative.
2. Label enough vials to provide several replicates of each tissue sample and fill the vials with fixatives. The African violet leaf

Table 27.1.   Graded Ethyl Alcohol
Dehydration Series and Immersion Times

Ethyl Alcohol Concentration (%)	Immersion Time (min)
10	30
20	30
30	30
40	30
50	30
60	30
70	60
80	60
90	60
100	60

sprouts, pea root tips, and callus tissue should be fixed in FAA solution; half of the onion roots should be fixed in FAA and half in Navashin's solution.

3. Place several pieces of each type of tissue into their respective vials and seal the vials.
4. Pipette off the fixative after 12 hr and quickly add fresh fixative. Since both fixatives are stable, the samples may be stored in the fixatives indefinitely.

Dehydration of Tissue
1. Prepare a graded ethyl alcohol series according to the protocol in Table 27.1. You will need about 500 ml of each solution.
2. Prepare a graded *tert*-butyl alcohol series according to the protocol in Table 27.2. You will need about 500 ml of each solution.
3. From each tissue fixation vial, pipette off the fixative, then add double-distilled water and gently swirl the tissue. Pipette off the

Table 27.2.   Graded *Tert*-Butyl Alcohol Dehydration Series and Immersion Times

Tert-Butyl Alcohol (ml)	Ethyl Alcohol (ml)	Water (ml)	Immersion Time (min)
10	20	70	30–60
15	25	60	30–60
25	30	45	30–60
40	30	30	Overnight
55	25	20	60–120
70	20	10	60–120
85	15	0	120–240
100	0	0	120–240

wash water and repeat the washing procedure 5 times over a 1-hr period.

4. Half of the callus specimens are to be dehydrated in the ethyl alcohol series and half in the butyl alcohol series. All other tissue specimens are to be dehydrated in the ethyl alcohol series.

5. Dehydrate each tissue specimen by immersing the tissue in either the graded ethyl alcohol or *tert*-butyl alcohol series, keeping the tissue in each concentration for the times indicated in Table 27.1 or 27.2. Add and remove solutions with Pasteur pipettes, making sure that the tissue specimens are not allowed to dry before the addition of the next solution.

Infiltration with Paraffin

1. After the required time in the cleaning agent, chloroform, butyl alcohol, or the last solution in the dehydration step, add a small amount of paraffin shavings or chips to each specimen vial, cork, and leave at room temperature for 4 hr.

2. When no more paraffin will go into solution, remove the cap from the vial, add more paraffin, and place the vial in an oven at 35°C for 12–24 hr.

3. Pour off 2/3 of any melted paraffin in the vial, add fresh melted paraffin, and place the vial at 55°C.

4. At intervals of 8–12 hr, pour off melted paraffin, replace it with fresh melted paraffin, and place vial at 55°C. Repeat the procedure at least three times.

Embedding

1. Collect several small 25 cm × 25 cm plastic weigh boats as these serve as excellent trays for embedding. Place weigh boats on a small hot plate heated to 45°–50°C.

2. Quickly remove a vial containing tissue from the oven and pour the contents into the weigh boat. If the amount of paraffin in the vial is insufficient to adequately cover the material, add more melted paraffin.

3. Arrange the specimens as desired with a hot needle or probe.

4. Carefully remove the weigh boat and sample from the hot plate and allow to cool.

5. Once the paraffin has cooled and hardened, the plastic weigh boat can be easily peeled away and reused. Place the paraffin block containing the tissue specimens in a labeled envelope.

Sectioning

1. Using a scalpel and a needle, carefully cut the paraffin block so one unit of tissue with about 0.5 cm of surrounding paraffin is

removed with the cut. Care must be taken to cut gently through the paraffin block because a quick, hard cut will shatter the entire block and possibly ruin several samples.

2. Obtain several metal stubs or mounting blocks. Gently heat the stub using an ethanol burner. With the other hand, press the paraffin block with the specimen onto the heated mounting stub.

3. Once the specimen is mounted onto the stub, run a heated needle between the mounting stub and specimen block several times in order to fuse the two. Additional paraffin may be melted around the base of the specimen block where it is in contact with the mounting stub.

4. Plunge the specimen and mounting stub into cool water to solidify.

5. Once the specimen block is firmly attached to the mounting stub, trim the specimen block using a single-edge razor blade to form a rectangle with parallel sides.

6. The specimen block is cut into 10-μm sections with a rotary microtome using a microtome block. Follow manufacturer's directions for sectioning, as methods vary with different instruments. Defects that may appear during sectioning and possible remedies are summarized in Table 27.3.

Affixing Sections to Slides

1. Place a small drop of egg albumin fixative on a clean microscope slide. With a clean finger, smear the drop evenly over the entire slide until there is a very thin layer of fixative on the slide.

2. Place a small drop of distilled water on the slide, which should be lying on a flat surface.

3. With a sharp scalpel, cut the paraffin ribbon obtained in the sectioning process into pieces of appropriate size (determined by the size of the coverslip).

4. Using a pair of forceps, pick up the paraffin ribbon and place it in the center of the water drop on the slide.

5. Place the slide on a hot plate sufficiently warm enough (35°–45°C) to expand the ribbon to its full size without melting the paraffin. Straighten the ribbon before complete expansion with a needle or probe.

6. When the ribbon has stretched completely, drain off excess water with a piece of tissue paper; carefully clean around the ribbon to remove all excess water and fixative. Place slide back on the hot plate for several hours.

7. When the slide is completely dry, remove it from the hot plate and place in a slide box or another dust-free area until you are ready to stain it.

8. Repeat steps 1–7 in the preparation of all slides.

Table 27.3. Possible Defects in Microtome Sections and Remedies

Defect	Possible Causes	Remedies
Ribbon curbed	1. Edges of block not parallel.	1. Trim block.
	2. Knife not uniformly sharp, causing more compression on one side than other.	2. Try another portion of knife-edge or re-sharpen knife.
	3. One side of block warmer than other.	3. Allow block to cool.
Sections compressed	1. Knife blunt.	1. Try another portion of knife-edge or re-sharpen knife. Compression often occurs through a rounded cutting facet produced by overstropping.
	2. Wax too soft at room temperature for sections of thickness required.	2. Reembed in suitable wax or cut thicker sections.
	3. Wax warmer than room temperature	3. Cool block at room temperature.
Sections alternately thick and thin, usually with compression of thin sections	1. Block or wax holding block to holder still warm from mounting.	1. Cool block and holder to room temperature.
	2. Block or wax holding block to holder cracked or loose.	2. Check all holding screws. Remove block from holder and holder microtome. Melt wax off holder, and make sure holder is dry. Recoat holder and remount block. Cool to room temperature.
	3. Knife loose.	3. Release all holding screws and check for dirt, grit, or soft wax. Check knife carriage for wax chips on bearing.
	4. Knife cracked.	4. Replace knife.
	5. Microtome faulty.	5. Return microtome to maker for overhauling.
Sections bulge in middle	1. Wax cool in center, warm on outside.	1. Allow block to adjust to room temperature. This is the frequent result of cooling blocks in ice water.

(*continued*)

Table 27.3. (Continued)

Defect	Possible Causes	Remedies
	2. Only sharp portion of knife is that which cuts center of block.	2. Try another portion of knife-edge or re-sharpen knife.
	3. Object impregnated with hard wax and embedded in soft wax, or some clearing agent remains in object.	3. Reembed object.
Object breaks away from wax or is shattered by knife	1. If object appears "chalky" and shatters under knife blade, it is not impregnated.	1. Throw block away and start again. If object is irreplaceable, try dissolving wax, rede-hydrating, reclearing, and reembedding.
	2. If object shatters under knife but is not chalky, it is too hard for wax sectioning.	2. Spray section between each cut with cal-loidin.
	3. If object pulls away from wax but does not shatter, the wrong dehydration process, clearing agent, or wax has been used.	3. Reembed in suitable medium.
Ribbon splits	1. Nick in blade of knife.	1. Try another portion of knife-edge.
	2. Grit in object.	2. Examine cut edge of block. If face is grooved to top, grit has probably been pushed out. Try another por-tion of knife-edge. If grit is still in place, dissect out with nee-dles. If there is much grit, throw block away.
Block lifts ribbon	1. Ribbon electrified. (Check by testing whether or not ribbon sticks to everything else.)	1. Increase room humid-ity. Ionize air with high-frequency dis-charge of bunsen flame a short distance from knife.
	2. No clearance angle.	2. Alter knife angle to give clearance angle.
	3. Upper edge of block has fragments of wax on it (a common result of there being no clearance angle).	3. Scrape upper surface of block with safety razor blade.

Table 27.3.   (Continued)

Defect	Possible Causes	Remedies
	4. Edge of knife (either front or back) has fragments of wax on it.	4. Clean knife with xylene.
No ribbon forms	1. Wax contaminated with clearing agent causing wax to crumble.	1. Reembed.
	2. Very hard, pure paraffin used for embedding so that sections, though individually perfect, do not adhere.	2. Dip block in soft wax or wax-rubber medium.
	3. a. Wax too hard at room temperature for sections of thickness required causing sections to roll into cylinders.	3. a. Reembed in suitable wax. If the section is cut very slowly, and the edge of the section held flat with a brush, ribbons may sometimes be formed.
	b. Knife angle wrong causing sections to roll into cylinders.	b. Adjust knife angle.

Staining with Chlorozol Black E

1. Prepare a 1% solution of Chlorozol Black E in 70% ethyl alcohol.
2. Prepare a series of containers large enough to dip slides into according to the following schedule:

Treatment	Immersion Time (min)
1. Xylene	15
2. Xylene and absolute ethyl alcohol (1 : 1)	10
3. Absolute ethyl alcohol	5
4. 95% ethyl alcohol	5
5. 70% ethyl alcohol	5
6. 1% Chlorozol Black E	5
7. 70% ethyl alcohol	2–3
8. 95% ethyl alcohol	5
9. Absolute ethyl alcohol	2–3
10. Xylene and absolute ethyl alcohol (1 : 1)	5
11. Xylene I	5
12. Xylene II	5

3. Stain each slide by immersing it in sequence in the solutions listed in the previous step for the indicated times. Refer to later section (Cementing Coverslips and Cleaning Slides) for final steps in slide preparation.

Staining with Hematoxylin and Aniline Blue

1. Prepare iron alum mordant containing 500 ml of 4% ferric ammonium sulfate solution, 5 ml glacial acetic acid, and 6 ml of 10% sulfuric acid.
2. To prepare 0.5% hematoxylin stain, add a pinch of sodium bicarbonate to 1 liter of distilled water. Bring the water to the boiling point, remove from heat, and add 5.0 g of the dye. **Do not boil the solution.** Cool and store in the refrigerator. Dilute the stock solution with 4 parts of water before use.
3. Prepare a saturated solution of aniline blue dye in methyl cellosolve.
4. Prepare a series of containers large enough to dip slides into according to the following schedule:

Treatment	Immersion Time (min)
1. Xylene	15
2. Xylene and absolute ethyl alcohol (1:1)	10
3. 95% ethyl alcohol	10
4. 70% ethyl alcohol	10
5. 50% ethyl alcohol	10
6. 35% ethyl alcohol	5
7. Rinse in distilled water	
8. Iron alum mordant	30
9. Rinse in distilled water	
10. 0.5% hematoxylin	12 hours
11. Rinse in distilled water	
12. 2% aqueous $FeCl_3$	3–5
13. Wash thoroughly under running tap water	30
14. 50% ethyl alcohol	5
15. 70% ethyl alcohol	5
16. 95% ethyl alcohol	5
17. Absolute ethyl alcohol	5
18. Absolute ethyl alcohol	5
19. Aniline blue stain	3
20. Rinse in absolute ethyl alcohol	

21.	Methyl cellosolve : xylene : absolute ethyl alcohol (1 : 1 : 2)	10–15
22.	Methyl cellosolve : xylene : absolute ethyl alcohol (2 : 1 : 1)	10–15
23.	Xylene and absolute ethyl alcohol (1 : 1)	5
24.	Xylene I	5
25.	Xylene II	5

5. Stain each slide according to the treatment schedule in the previous step. Immerse slides for the indicated times or rinse as indicated. Refer to later section (Cementing Coverslips and Cleaning Slides) for final steps in slide preparation.

Staining with Safranin and Fast Green

1. Prepare a 1% safranin solution in water and a saturated solution of fast green in absolute alcohol.
2. Prepare a series of containers large enough to dip slides into according to the following schedule:

Treatment	**Immersion Time (min)**
1. Xylene I	1–3
2. Xylene II	2
3. Absolute alcohol	2–5
4. 70% ethyl alcohol	2–5
5. 30% ethyl alcohol	2–5
6. Distilled water	1–2
7. 1% safranin	5
8. Tap water, rinse until excess is removed	
9. 50% ethyl alcohol	6
10. 90% ethyl alcohol	5
11. Absolute alcohol	2
12. Fast green in absolute alcohol	1–3
13. Absolute alcohol : xylene (1 : 1)	3
14. Absolute alcohol : xylene (1 : 9)	3
15. Xylene I	3–5
16. Xylene II	5

3. Stain each slide according to the treatment schedule in the previous step by immersing the slide for the indicated times. Complete preparation of slides as described in the next section.

Cementing Coverslips and Cleaning Slides

After removing a slide from the last container of pure xylene in the staining schedule, wipe off the excess xylene from the back of the slide and around the specimen. Lay the slide on a level surface and place several small drops of Permount (Fisher Sci.) on top of the specimen. Gently lower a coverslip over the specimen, spreading the Permount evenly in order to avoid air bubbles. Allow the slide to dry at room temperature for 24 hr on a level surface. After the coverslip has dried, clean the slide with xylene. Care must be taken not to use too much xylene and uncement the coverslip.

Interpretation of Stains
1. Chlorozol Black E stains nuclei and celluose cell walls.
2. Hematoxylin stains cellulose cell walls, middle lamella, chromosomes, and mitochondria.
3. Aniline blue stains cellulose cell walls.
4. Safranin stains lignified cell walls, cutinized cell walls, and chromosomes.
5. Fast green stains cellulose cell walls and cytoplasm.

REFERENCES

1. Berlyn, G. P. and J. P. Miksche. 1976. Botanical Microtechnique and Cytochemistry. p. 326. Iowa State Univ. Press, Ames.
2. Clark, G. 1973. Staining Procedures Used by the Biological Stain Commission. Williams and Wilkins, Baltimore.
3. Conn, H. J. 1960. Biological Stains. Williams and Wilkins, Baltimore.

28

Specimen Preparation for Scanning Electron Microscopy

Scanning electron microscopy (SEM) has become a useful tool for examining the external characteristics related to various developmental processes in plants. With SEM, the external fractures of a tissue can be observed with a greater depth of field and at a higher magnification and resolution than with light microscopy.

There is little argument that the most critical part of SEM is adequate preparation of the specimen. The procedures can vary greatly depending upon the specimen and generally are most complex for biological tissues. In preparing biological specimens for SEM, consideration must be given to the possible presence of natural materials on the specimen surface or surface contaminants, which will distort the specimen. Specimens may require a gentle washing either before or during the fixation process to remove surface contaminants. Care must be taken to chose a buffer and/or washing solution with an appropriate pH and osmolarity. Specimen preparation for SEM generally involves three major steps: fixation, dehydration, and critical point drying.

Fixation is probably the most critical step in the preparation of biological tissue (5). The purpose of fixation is to preserve the fine structure of the cells and to make them stable to alterations or distortions that may occur during subsequent treatments. A properly fixed specimen should be as representative as possible of the living state. Specimen tissue may be fixed in one of two ways. One method—mechanical fixation—relies upon the rapid freezing of the tissue to maintain its structural stability. The second method and more commonly used method is chemical fixation, which relies upon chemicals to kill and fix the tissue.

There are two types of chemical fixatives: coagulative fixatives and noncoagulative fixatives. Coagulative fixatives were developed first and are the mainstay of light microscopy. This group of fixatives includes formalin–acetic acid–alcohol (FAA), chromic acid–acetic acid–formalin (CRAF or Navaschin), and several others. In tissues treated with these chemicals, the cellular contents, especially the proteins, become coagulated and the integrity of the nucleoplasm and cytoplasm is distorted (4). Noncoagulative fixatives do not disrupt the cellular structure of tissues as much as coagulative fixatives. Some noncoagulative fixatives are glutaraldehyde, paraformaldehyde, osmium tetroxide, and acrolein. This group of fixatives forms molecular cross-links, which provide structural stability without causing coagulation.

A suitable cellular environment must be present during fixation if cellular structures are to be maintained. Therefore, the osmolarity, temperature, and pH of the buffer and/or fixative are critical variables during the fixation process. The osmolarity of the fixative for SEM should be somewhat more isotonic than that of the living tissue in order to avoid any osmotic shock to the surface layers of the specimen, which are the areas of interest. The temperature and pH should also be maintained relatively close to normal conditions to ensure proper membrane and cellular integrity. The pH for plant tissue fixation should be maintained around 7.1–7.2 with a suitable buffer (5).

Fixatives are applied to plant tissue either by vapor fixation or by immersion. In vapor fixation, the specimen is in a vial in close proximity to an open vial of the fixative; both vials are confined within a closed container. Another method is to suspend the specimen in a nylon net or mesh directly above the surface of the fixative contained in a sealed vial. The specimen eventually becomes fixed by the vapors of the fixative. The fixative usually employed in vapor fixation is osmium tetroxide. This method of fixation is excellent for delicate specimens such as fungi. The other method of applying a fixative is by immersion. In this method, the specimen must be dissected into small enough pieces so that the fixative will penetrate through layers of the tissue.

The second major step in specimen preparation is dehydration. Most tissues must be dehydrated to remove water from the specimen before inserting it into the microscope, which is under vacuum. Improper dehydration of the tissue and water loss can result in contamination of the column and damage to the topography of the specimen. Chemicals commonly used to dehydrate tissues include organic solvents such as ethanol, acetone, dioxane, ether, chloroform, and amylacetate (1, 2, 3). Dehydration is generally accomplished by passing the specimen through a series of graded concentrations of the solvent, which gradually removes the water from the tissue and reduces the surface tension. The time spent in each solution varies with the type of tissue and the solvent, but sufficient time to allow penetration of each solution in the series into the specimen must be allowed. Ethanol and acetone are the most commonly used solvents for dehydrating tissues for SEM.

The third step in specimen preparation is critical point drying. In this process, liquids change from the liquid phase to the gaseous phase without the heat of vaporization occurring. The temperature and pressure at which this occurs is called the *critical point*. This phenomenon results in the removal of the organic solvents from the tissue, producing a dry specimen free of surface distortions. The procedures described in this chapter for the preparation of specimens for SEM have been used successfully with a variety of materials including delicate specimens such as fungal mycelium. The procedure described for fixation, while possibly tedious and lengthy, has proven quite successful for soft, delicate biological specimens such as those encountered in tissue culture.

Materials Required
1.   20 sterile plastic or glass petri plates
2.   2 scalpel handles and blades
3.   20 small specimen vials (10–15 ml) with closures
4.   Stock solutions of 0.1 M KH_2PO_4 (13.6 g/liter) and 0.1 M Na_2HPO_4 (14.19 g/liter) in double-distilled water
5.   Phosphate buffer of the required pH prepared by mixing stock solutions of KH_2PO_4 and Na_2HPO_4 in the following ratios (v/v):

pH	0.1 M KH_2PO_4	0.1 M Na_2HPO_4
6.33	76	28
6.41	68	32
6.53	62	38
6.61	56	44
6.70	52	48
6.81	48	52
6.91	40	60
7.00	34	66
7.10	28	72
7.24	22	78
7.30	20	80
7.42	16	84
7.57	12	88

(After mixing solutions, check pH with a pH meter and, if necessary, adjust to desired pH with NaOH or HCl.)
6.   Specimen stubs
7.   Double-stick tape for mounting specimens
8.   Dissecting microscope

Preparation of Fixatives

1. Glutaraldehyde is generally purchased in highly purified liquid form. To prepare glutaraldehyde fixative solution, mix an appropriate volume of 25% or 50% commercial glutaraldehyde with enough phosphate buffer of the desired pH and molarity to yield a final buffered glutaraldehyde concentration of 1.5–6%. The lower concentrations are generally more desirable, since the higher concentrations may result in tissue shrinkage. Concentrations of 1.5–3.0% have worked well with plant tissue cultures.

2. Osmium tetroxide is a highly dangerous fixative, which must be prepared under a fume hood. Osmium tetroxide can be obtained from commercial suppliers in both the liquid and crystalline form. The liquid form is generally available in 5% aqueous solutions and is added directly to an appropriate volume of buffer to give the final desired concentration (usually 1%). To prepare a fixative solution starting with crystalline osmium tetroxide, proceed as follows:

    a. Gently heat the ampule containing crystalline osmium tetroxide under running hot water. This will accelerate the osmium going into solution with the buffer.
    b. Clean the ampule with acetone and place it in a glass-stoppered bottle.
    c. Under a fume hood, break open the vial and add enough buffer of the desired pH to give a 1% solution.
    d. Tightly seal the container and place it in a refrigerator for at least 24 hr before using the solution. The fresh solution should be light amber in color and may be stored for several weeks at 4°C.

3. Prepare a saturated thiocarbohydrazide solution (TCH) by washing 0.5 g TCH with double-distilled water until all coloration is removed. Then heat the rinsed TCH in 25 ml of double-distilled water to 60°C and hold for several minutes. Cool the solution to 25°C and allow the solution to sit for 1 hr. Finally, filter the saturated solution through a 0.4 μm Millipore filter. Prepare fresh solution for each group of specimens.

Fixation

1. Examine cultures carefully under a dissecting microscope and carefully excise the area of interest to be examined. Unnecessary portions of the specimen should be trimmed away, since only a small portion of tissue can be placed in the microscope.

2. Place each specimen in a small vial (10–15 ml) and gently add

enough cold (4°C) buffered glutaraldehyde to cover the sample two to three times. Place vials in a refrigerator at 4°C for 12 hr.

3. Decant the buffered glutaraldehyde and gently wash each specimen several times with cold (4°C) buffer.

4. After removing the final buffer wash, add enough cold buffered osmium tetroxide solution to cover specimen two to three times. Store specimens at 4°C for 12 hr.

5. Decant osmium tetroxide and rinse tissue with cold (4°C) double-distilled water several times. After decanting the final water rinse, gently add enough filtered thiocarbohydrazide solution to cover the sample two to three times. Leave specimens in thiocarbohydrazide for 60 min at 4°C.

6. Remove all thiocarbohydrazide and then wash each sample several times in cold double-distilled water to remove any unbound thiocarbohydrazide. After removing the last rinse, add enough buffered 1% osmium tetroxide to completely cover sample. Store vials at room temperature for 60 min.

7. Remove osmium tetroxide from each sample vial and wash the specimen several times with double-distilled water. Fixed specimens in double-distilled water may be stored at 25°C for several months before dehydration.

Dehydration

1. Dilute appropriate amounts of acetone or ethyl alcohol with enough double-distilled water to give 100 ml of each of the following concentrations: 20, 40, 60, 80, 90, 95, 100, and 100%.

2. Decant the water from each specimen vial and replace with enough 20% ethanol or acetone solution to cover the specimen two or three times. Leave the specimen in this solution for 30–60 min depending upon the size of the specimen (the larger the specimen the longer it will take for the organic solvent to penetrate).

3. Decant the 20% ethanol or acetone solution and replace with the next solution (40%) in the dehydration series. Repeat this process for the remaining concentrations in the dehydration series listed in step 1, holding the specimen in each solution for 30–60 min.

4. The specimen can be held indefinitely in a tighly sealed container at 100% ethanol or acetone. Store the specimen carefully to prevent damage to the specimen surface due to sudden movement.

Critical Point Drying

The instructions in this section are for a Denton Critical Point Dryer (Fig. 28.1). However, the procedure for using most other critical point dryers is similar. Liquid carbon dioxide (CO_2) is used as the transitional

Fig. 28.1 Denton Critical Point Dryer Functional Components. 1) CO_2 tank valve. 2) Specimen chamber cap. 3) Specimen chamber. 4) Pressure gauge. 5) CO_2 inlet valve. 6) CO_2 exhaust valve. 7) Beaker. 8) Beaker support tray.

fluid since it has the most acceptable critical temperature ($T_c = 31°C$) and pressure ($P_c = 1072$ psi) for plant specimens.

1. Place a small amount of dehydrant (100% acetone or ethyl alcohol) into a glass petri dish (volume of dehydrant must be such that the sample is completely immersed). Remove cap from a wire mesh specimen basket, place the basket in the dish, and transfer the dehydrated sample from vial into the basket. Replace basket cap. Repeat for each specimen.

2. Unscrew and remove the pressure chamber cap and place an appropriate amount of dehydrant into the chamber. *Quickly* transfer the specimen baskets into the chamber and replace the cap to seal the chamber.

3. After insuring that *both* inlet and exhaust valves are closed, slowly open the CO_2 tank valve. Then open the inlet valve very slowly to admit liquid CO_2 into the chamber. The pressure gauge will read both tank and chamber pressure. The gauge should rise slowly; after about 1 min, when the full tank pressure is reached (700–900 lb), open the inlet valve completely. Fill the stainless steel beaker with cold water to within 2 in of the top and immerse

the pressure chamber into it. Swing the beaker tray into the supportive position.

4. Slowly open the exhaust valve being careful not to let the pressure within the chamber fall (monitor pressure gauge). Observe the exhaust to monitor flow. The flow should be as low as possible for minimum distortion of the specimens. Flush until the major portion of the dehydrant is removed and a good flow of solid CO_2 is coming out the exhaust. Continue to flush for 1 min.

5. Close the exhaust valve and leave it closed for 10 min; then open the exhaust valve for 6 min. Repeat this flushing operation 5 more times. When the 4th flushing period is started, turn on the hot water tap and let the water run. The water should then be hot enough (50°–60°C) when it is required.

6. After the 6th flush, close the (1) exhaust valve, (2) inlet valve, and (3) CO_2 tank valve in that order. Fill the stainless steel beaker with hot water to within 2 in of the top and immerse the pressure chamber in it. The pressure will begin to rise in the chamber (monitor on the pressure gauge); whenever the pressure stops rising, change the water. After the pressure reaches 1600 psi, the CO_2 is above the critical pressure and temperature. Do not allow the pressure to exceed 2000 psi.

7. Open the exhaust valve slowly and gradually bleed off the gas to reduce the pressure in the chamber to 1 atmosphere. The exhaust rate should be 2–3 psi/sec; it should take about 10 min to completely remove the gas. A fast rate of bleeding will result in specimen damage.

8. When the pressure in the chamber reaches zero, remove the beaker and dry off the pressure chamber with a towel. Unscrew the pressure chamber cap and quickly remove the specimen baskets. Immediately remove the specimens from the baskets and place them in a chambered petri dish that contains desiccant in one of the chambers. Petri plates containing dried specimens should be stored in a desiccator containing active desiccant. Handle the dried specimens with extreme care because they are very light, delicate, and brittle.

Mounting Specimens

Aluminum or brass stubs can be used for mounting of the specimen. The specimen must be securely mounted to the stub by an adhesive material, which is also electrically conductive. Two of the more common adhesives are liquid carbon black and double-stick tape. A dissecting microscope may be required to properly align the specimen onto the stub.

Sputter-Coating the Specimen

The final step before a specimen can be observed with the SEM is the coating of the specimen with a thin layer of a conductive material such as carbon, gold, or gold–palladium. The coating material conducts heat and electrical charges away from the specimen to the specimen stub, avoiding charging or beam damage artifacts. The principle behind most sputter coaters is the same. The instructions in this section are for a Polaron PE-5000 sputter coater (Fig. 28.2).

1. Open argon tank valve.
2. Open specimen chamber lid, place specimen stubs on specimen stage, and close the specimen chamber lid.
3. Set timer to 2 min by depressing the triangular knob and rotating either clockwise or counterclockwise.
4. Set operation knob to "PUMP" position and evacuate chamber until the Pirani gauge reads 0.1–0.09 torr.
5. Open argon leak valve by rotating it counterclockwise one complete revolution from the zero position; quickly close it again by rotating it clockwise back to the zero position. Allow the specimen chamber to evacuate until the Pirani gauge reaches 0.07 torr or less. Repeat this procedure another four to six times. This will ensure that the residual gas in the chamber is essentially all argon.

Fig. 28.2   Polaron PE-5000 Sputter Coater.

6. After the last flushing procedure and *when the vacuum is 0.07 torr or less,* set operation knob to the "SET HT" position. Rotate the high-tension knob clockwise until the black line coincides with 1.2 kV.

7. Slowly open the argon leak valve by rotating it counterclockwise until the milliammeter needle reads 40. A blue-violet discharge will be evident in the specimen chamber.

8. Turn operation knob to the "TIMER" position and activate the timer by pressing the black push button. Maintain a reading of 40 on the milliammeter by minor adjustments of the argon leak valve. If the reading drops below 40, open the leak valve a little more; if the reading rises above 40, close the leak valve a little more.

9. When the timer goes off, do the following in the order listed:

    a. Rotate the high-tension knob counterclockwise to its stop position. This turns off the high tension.
    b. Rotate the operation knob to the "OFF" position.
    c. Admit air gradually to the specimen chamber by lifting the air admittance valve on the specimen chamber lid.
    d. Close the argon leak valve. **Do not exceed zero!**
    e. Close the argon tank valve.
    f. Remove specimens and close the specimen lid.

REFERENCES

1. Boyde, A. and C. Wood. 1969. Preparation of animal tissues for surface scanning electron microscopy. J. Microscopy 90:221–249.
2. Humphreys, W. J. 1975. Drying soft biological tissues for scanning electron microscopy. pp. 707–714. *In* SEM/1975. IIT Research Institute, Chicago, IL.
3. Marszalik, D. S. and E. B. Small. 1969. Preparation of soft biological materials for scanning electron microscopy. pp. 231–239. *In* SEM/1969. IIT Research Institute, Chicago.
4. O'Brien, T. P., J. Kew, M. E. McCully, and S. Y. Zee. 1973. Coagulant and non-coagulant fixation of plant cells. Aust. J. Biol. Sci. 26:1231–1250.
5. Postek, M. T., K. S. Howard, A. H. Johnson, and K. L. McMichael. 1980. Scanning Electron Microscopy. Ladd Research Industries, Inc.,

Appendix 1: Interconversion of Centigrade and Fahrenheit Temperature Scales

C°	F°	C°	F°	C°	F°
−18	0	16	61	35	95
−10	14	17	63	36	97
−5	23	18	64	40	104
0	32	19	66	50	122
1	32	19	66	50	122
2	34	20	68	60	140
3	37	22	72	80	176
4	39	23	73	90	194
5	41	24	75	100	212[a]
6	43	25	77	121	250[b]
7	45	26	79	160	320[c]
8	46	27	81		
9	48	28	82		
10	50	29	84		
11	52	30	86		
12	54	31	88		
13	55	32	90		
14	57	33	91		
15	59	34	93		

For precise conversions use the following formulas:

$$F° = \tfrac{9}{5} C° + 32°$$
$$C° = \tfrac{5}{9} (F° - 32°)$$

[a] Boiling point of water at sea level
[b] Autoclaving temperature
[c] Dry heat sterilizing temperature

Appendix 2: Approximate Interconversion of Some Common Units of Measure

LENGTH

1 inch = 2.54 centimeters
1 centimeter = 0.4 inches

WEIGHT

1 ounce = 28 grams
1 gram = 0.036 ounces
1 pound = 0.45 kilograms
1 kilogram = 2.2 pounds

VOLUME

1 quart = 0.95 liters
1 liter = 1.06 quarts
1 fluid ounce = 30 milliliters
1 teaspoon = 5 milliliters
1 tablespoon = 3 teaspoons
1 cup = 16 tablespoons

LIGHT

1 foot candle = 10.8 lux
1 lux = 0.093 foot candles

Conversions from foot candle or lux (illuminance) to absolute units of irradiance can only be approximate unless detailed measurements are made. The following roughly approximate conversion factors are useful:

Incandescent light (PAR range): 1 foot candle = 4.5 microwatts/cm^2
Cool white fluorescent light (PAR range): 1 foot candle = 3.5 microwatts/cm^2

Appendix 3: Washing and Handling Glassware

The glassware that tissue or cell cultures come in contact with, either directly or indirectly via the medium, must be free of contaminating substances. Because glassware often is contaminated or can easily become contaminated, it is extremely important that all tissue culture glassware be handled with extreme care and its treatment closely supervised. Many detergents for tissue-culture glassware are currently marketed and are generally satisfactory; however, the cleaning and rinsing procedures must be adjusted to each particular compound. The following procedures have been employed in a number of tissue-culture facilities and may be helpful as general guidelines.

GENERAL PROCEDURES

1. All reusable tissue-culture glassware should be emptied and soaked immediately after use. Media should not be allowed to dry on the glassware.
2. All glassware that has contained corrosive chemicals or fixatives should be segregated from the rest of the tissue-culture glassware.
3. All glassware containing or coming in contact with microorganisms should be decontaminated before washing.
4. Labels and marking ink should be removed before washing. Marking ink may be removed by wiping with an abrasive cleanser or acetone and then immediately rinsing.
5. Vessels containing agar should be autoclaved or filled with hot water to melt the agar before washing. The agar should be poured from the vessel into a collecting sieve and discarded.
6. All glassware that has been treated with silicone should be permanently labeled and segregated from the rest of the glassware.
7. Rubber-lined screw caps should never be soaked in a detergent or soap solution. Soak caps only in distilled water.
8. Glassware used for growing cells should be acid-cleaned if proteinaceous deposits cannot be removed by conventional washing (see later section for instructions).

GLASS-WASHING PROCEDURES

1. Tissue-culture glassware can be soaked in a 5% detergent solution for a minimum of 1 hr.
2. Following detergent washing, the glassware is brushed to remove any residual matter and is then placed into an automatic washer. The automatic washer is used only as a rinsing unit and *no detergent is added to the washer.*
3. The automatic washer is programmed for no less than six tap water rinses and a 1-min distilled-water rinse.
4. If a piece of glassware does not fit into an automatic unit, it is rinsed by hand. Hand rinsing requires a minimum of six thorough tap water rinses and four rinses in quality water.
5. After rinsing, glassware should be dried thoroughly. Glassware can be dried at 70°C or be covered and air-dried overnight.

ACID-CLEANING PROCEDURE

1. All personnel should wear a full face mask, full-length acid-resistant apron, and acid-resistant gloves.
2. Glassware to be acid-cleaned should be thoroughly prerinsed with water before acid cleaning.
3. Handling of acid and processing of glassware should be done in a fume hood.
4. A sulfuric acid solution containing potassium dichromate is effective for acid treatment of glassware. **Caution: Never add water to acid.** If it is necessary to dilute acid, slowly add acid to water while stirring.
5. Glassware is placed in freshly prepared acid bath for at least 1 hr. When the cleaning solution becomes dark brown, it should be discarded.
6. Glassware is removed from solution and excess acid drained off. Glassware is then placed in a suitable container of distilled water, removed from the hood, and thoroughly rinsed before normal detergent washing.

Appendix 4: Preparation of Culture Media

It is extremely important, when preparing a culture medium, that all weighing be done with the utmost precision and that the purest of chemicals be used. The omission of one component or the addition of the incorrect amount can have disastrous consequences. Use of prepared media, which are formulated precisely, is convenient and often advisable. However, commercially prepared media are only available in a limited number of specific formulations, none of which may meet your needs. Observe the following general procedures when preparing culture media:

1. Unless specifically instructed otherwise, use Reagent grade chemicals.
2. Follow, step by step, the protocol for the preparation of the specific medium or stock solution being made.
3. As each step is performed, record the exact weights or volumes of the reagents used and the source from which these were drawn on standard Log for Media and Solution Preparation forms. Use one form for each solution or medium prepared.
4. When you remove an aliquot from any stock solution, record the date, amount used, your initials, and amount remaining on the stock solution bottle.
5. Dispense liquid media in the required amounts into culture vessels with a graduated cylinder or syringe and close the vessels with appropriate plugs, closures, screws, caps, etc., and autoclave.
6. Before dispensing agar-containing media, add the required amount of agar, suspend, and heat on a hot plate/stirrer to 210°F for 5 min. Then pour media into culture vessels, close them, and autoclave.
7. If the medium contains heat-labile components, filter-sterilize these separately, then add them to the rest of the medium after it has been autoclaved and cooled. Dispense aseptically into sterile culture vessels.

8. Autoclave media as follows:

   a. Volumes of less than 200 ml—15 min at 121°C and 15 psi.
   b. Volumes of 200–1000 ml—30 min at 121°C and 15 psi.
   c. Volumes larger than 1 liter—obtain specific instructions.

9. After autoclaving, remove media as soon as the autoclave temperature reaches 100°C and the pressure within the autoclave is zero.

10. Cool, label, and tighten screw caps.

Appendix 5: Stock Solution Dilution Chart

Use this stock dilution chart to quickly determine the volume of a stock solution of a particular concentration (in mg/1000 ml) to use in preparing a given volume of a diluted final solution of a specified concentration. For example, assume you have a 25% stock solution (25 mg/100 ml) and wish to prepare 1 liter of final solution containing 0.125 mg/liter. In the left-hand column, find the stock solution containing 25 mg/100 ml and then read across the chart to the column headed 1 liter of final solution. Of the four values listed, 0.125 mg corresponds to the desired final concentration. Looking left to the amount to use column, you see that 0.5 ml of stock solution (25 mg/100 ml) is required to prepare 1 liter of final solution with a concentration of 0.125 mg/1000 ml.

Concentration of Stock Solution (mg/100 ml)	Amount to Use (ml)	Concentration of Final Solution (mg)			
		250 ml	500 ml	1 liter	2 liters
1	0.1	0.0004	0.0002	0.0001	.00005
	0.5	0.002	0.001	0.0005	.00025
	1.0	0.004	0.002	0.001	.0005
	10.0	0.04	0.02	0.01	.005
10	0.1	0.04	0.02	0.01	0.0005
	0.5	0.2	0.1	0.05	0.025
	1.0	0.4	0.2	0.1	0.05
	10.0	4.0	2.0	1.0	0.5
20	0.1	0.08	0.04	0.02	0.01
	0.5	0.04	0.02	0.1	0.05
	1.0	0.8	0.4	0.2	0.1
	10.0	8.0	4.0	2.0	1.0
25	0.1	0.1	0.05	0.025	0.05
	0.5	0.5	0.25	0.125	0.25
	1.0	1.0	0.5	0.25	0.5
	10.0	10.0	5.0	2.5	1.25

Concentration of Stock Solution (mg/100 ml)	Amount to Use (ml)	Concentration of Final Solution (mg)			
		250 ml	500 ml	1 liter	2 liters
50	0.1	0.2	0.1	0.05	0.025
	0.2	0.4	0.2	0.1	0.05
	0.5	1.0	0.5	0.25	0.125
	1.0	2.0	1.0	0.5	0.25
	2.0	4.0	2.0	1.0	0.5
	5.0	10.0	5.0	2.5	1.25
	10.0	20.0	10.0	5.0	2.5
100	0.1	0.4	0.2	0.1	0.05
	0.5	2.0	1.0	0.5	0.25
	1.0	4.0	2.0	1.0	0.5
	10.0	40.0	20.0	10.0	5.0

Appendix 6: Storage of Auxins, Cytokinins, Vitamins, and Enzymes

Compound	Storage Requirements (Dry Powder/ Stock Solution)	Supplier[a]
Auxins		
Indoleacetic acid (IAA)	Desiccate, −20°C	A
Naphthyleneacetic acid (NAA)	4°C	A
2,4-Dichlorophenoxyacetic acid (2,4-D)	4°C	A
Indolebutyric acid (IBA)	4°C	A
Cytokinins		
6-Furfurylaminopurine (kinetin)	Desiccate, −20°C	A
6-Benzyladenine (BA)	Desiccate, −20°C	A
6-(γ-γ-Dimethylallylamino) purine (2iP)	Desiccate, −20°C	A
6-(4-Hydroxy-3-methyl-*trans*-2-butenylamino)purine (zeatin)	Desiccate, −20°C	A
Vitamins		
Thiamin HCl	4°C in dark bottle	A
Nicotinic acid	4°C or room temp.	A
Folic acid	4°C or room temp.	A
Pyridoxine HCl	4°C or room temp.	A
Myo-Inositol	4°C or room temp.	A
Enzymes		
Cellulase	Desiccate, −20°C	A
Cellulysin	Desiccate, −20°C	B
Macerase	Desiccate, −20°C	B
Pectinase	4°C in the dark	A

[a] A = Sigma Chemical, P.O. Box 14508, St. Louis, MO 63178; phone 1(800)325-3010 and 1(314)771-5750.

B = Calbiochem-Behring Corp., P.O. Box 12087, San Diego, CA 92112; phone 1(800)854-2171 and 1(714)453-7331.

Appendix 7:   Desalting Commercial Enzyme Preparations

Many commercial enzyme preparations contain high concentrations of salt to stabilize the enzyme protein. Before such preparations are used in tissue culture work (e.g., in isolation of protoplasts), they may have to be desalted. The following general procedure can be used to desalt enzyme preparations:

1. Construct a Sephadex G-20 column, approximately 4.5 cm × 48.5 cm, and equilibrate with double-distilled water.
2. Dissolve 10.0 g of the commercial enzyme preparation in 100 ml of double-distilled water. Stir the enzyme solution very slowly at room temperature (75°C) for 1 hr.
3. Pour the enzyme solution into a centrifuge tube and centrifuge at 8000g for 15 min. The insoluble materials in the enzyme mixture will form a pellet in the bottom of the tube. Remove the supernatant and load onto the Sephadex column.
4. The enzyme solution will form light to dark brown regions on the column. Begin testing for protein content as soon as the first light brown fraction is collected, as follows:

   a. Mix 1 drop of the collected fraction with 2 drops of cold (4°C) 10% trichloroacetic acid (TCA). If protein is present, the sample will begin to foam or become cloudy.
   b. Test every fifth fraction until a negative reaction occurs.

5. Combine all protein-containing fractions in a glass petri dish and freeze-dry (lyophilize) for 15–16 hr. Store the lyophilized enzyme at −20°C until needed.

Appendix 8:   Sterile Filtration

For a number of reasons, membrane filtration under positive pressure is the method of choice for sterilization of culture media that contain heat-labile components and/or materials (e.g., animal sera or enzymes) that are subject to foaming and consequent surface denaturation if filtered under reduced pressure. During membrane filtration, the pH can be controlled within narrow limits by utilizing an appropriate mixture of CO_2 in air (usually 5%) for applying pressure to solutions buffered with bicarbonate. Membrane filters and filter holders are manufactured by Millipore Filter Corp., and the Gelman Co.

The final, smallest pore-size filter (0.22 μm) recommended in this procedure will retain bacteria, yeast, and molds. Since most viruses will pass through, autoclavable media should be chosen for virus studies whenever possible. The capability of a 0.22-μm filter to retain mycoplasma (about 0.15 μm in size) is uncertain. However, the titer of a mycoplasm-containing serum is said to be reduced by three or more orders of magnitude by filtering. Pleomorphic forms of bacteria may also be able to pass through 0.22-μm filters.

Some general considerations related to membrane filtration are discussed below. Operation of Seitz filters, which can be used to filter-sterilize media, is summarized in Appendix 17.

CHOICE OF APPARATUS

Several types and sizes of filters are available. Choice is governed by the volume of fluid to be filtered and convenience. For small volumes (10–100 ml), filters, available in three sizes from Millipore, are suitable. These can be attached to a syringe containing the fluid, and the positive pressure can be applied with the syringe. For volumes from 50 ml to 1 liter, stainless steel 47-mm pressure filter holders are available. These accommodate standard commercially available membrane filters. If more than 1 liter of fluid is to be filtered, stainless 142-mm filter holders are most efficient; these also can accommodate standard membrane filters of various porosities.

Nalgene, Corning and Bioquest market membrane filter units that are inexpensive enough to use once and then discard. They come completely assembled, sterilized, and individually sealed. Usually only a small volume, up to 500 ml, can be handled, however, and filtration is by negative pressure, which can cause foaming.

ASSEMBLY AND STERILIZATION

The manufacturer's instructions for assembling a specific filter should be referred to. To facilitate passage of viscous materials such as serum and embryo extract, the material can be filtered sequentially through a series of fiberglass prefilters of graded pore sizes, typically 0.8, 0.45, and 0.22 μm. Dacron separators are interposed between the 0.8- and 0.45- and between the 0.45- and 0.22-μm filters. Procedures for wrapping and connecting the filter to a receiving vessel will be demonstrated.

Filtration units are autoclaved assembled with the input and output parts covered with aluminum foil. To protect the membrane, the manufacturer's instructions for time, temperature, and pressure of autoclaving must be followed.

PRETREATMENT

Most membrane filters contain some type of detergent which is incorporated during manufacturing. Unless this is removed before the unit is used to filter media, it can be toxic to cells. To pretreat a detergent-containing filter, distilled water heated to approximately 80°C is passed through the filter, followed by a cold saline rinse. The filter is then ready for use. Nondetergent filters are commercially available from Gelman.

FILTRATION

Pressure from a cylinder of 5% CO_2 in air is applied gently to begin filtration. Minimum pressure for the desired flow rate should be utilized. The pressure should never exceed 15 psi with membrane filters. Fill storage bottles or tubes from sterilized bell apparatus (a bell or dome-shaped glass with a dispensing outlet inside the dome).

CLEANING

Membrane filters should never be allowed to dry dirty. After use, filters should be rinsed well and placed in a 1% detergent solution. Manufacturer's instructions should be consulted for detailed cleaning procedures.

Appendix 9: Determination of Cell Number in Cell Suspension and Protoplast Cultures

Transfer the cell pellet obtained in the determination of the packed cell volume or a known volume of a cell suspension to a 10-ml glass vial with a lid and add 2 ml of 10% HCl. Gently disturb the cell pellet and swirl to ensure adequate mixing of the cells with the acid, cover, and put in cold storage at −5°C for 16–24 hr. Remove the vial from cold storage and thaw at room temperature (22°C); then add 2 ml of 10% chromic acid and incubate at 22°–25°C for 3–5 days. Pour the cell suspension into a 10-ml graduated cylinder and bring to a known volume with a 1 : 1 mixture of 10% HCl and 10% chromic acid. Macerate any large cell clumps by drawing the suspension in and out of a Pasteur pipette or syringe five to ten times. Finally, vortex the suspension on a vortex mixer. Further dilution may be necessary if the cell count on a haemocytometer exceeds 250 per field. Resuspend the suspension in a 5–10% saline solution prior to counting. The above procedures are generally not required for protoplast cultures.

USE OF THE HAEMOCYTOMETER

The most useful haemocytometer for counting plant cells is one which has a clearance of 0.2 mm, because of the relatively large size of plant cells and their shape. A haemocytometer has two to four chambers, the depth of which is controlled by a coverglass 0.2 mm above the ruled areas. The chambers have rulings on them, the main outlines of which are indicated by triple lines. Each of the squares bounded by adjacent triple lines is 1 mm on a side. These 1 mm squares are arranged in 5 rows of 2. Each of the squares, therefore, represents a volume of 0.2 mm^3. Multiplication of the count per square by 5000 gives the number of cells or particles per milliliter.

Haemocytometer counts are subject to many errors. Critical attention must be given to obtaining a uniform suspension of cells or particles; to correct loading of the chambers; to adopting a standard convention for counting cells in contact with either boundary lines or other cells, etc. With the best techniques in the hands of an experienced worker, the error rate is likely to be 10–15%. A frequently ignored source of error is the standard error related to the total number of cells counted. In counting cells, the standard error is given by the square root of the number of cells counted because the distribution of cells is Poisson rather than Gaussian. Thus, for a count of 100 cells, the standard error is 10 or 10% of the total; for a count of 400 cells, the standard error is 20 or 5% of the total counted.

Before making a cell count, obtain the following materials: (1) clean haemocytometer and coverglass, (2) Pasteur pipettes, (3) compound microscope, preferably with a mechanical stage, and (4) hand tally counter. After preparing the cell suspension as described above, proceed as follows:

1. Seat the coverglass firmly over the haemocytometer chambers.
2. With a Pasteur pipette, quickly fill each chamber with cell suspension. Fill the chambers without overfilling by capillary action. Refill the pipette between each chamber filling. Agitate the suspension thoroughly before filling the pipette.
3. Using a microscope with a 10× ocular and 10× objective, count cells in the four chambers of the haemocytometer until a sufficient number of particles have been counted. If the reproducibility of the count is inadequate, repeat with fresh samples.
4. Clean the haemocytometer and coverglass by dipping into a detergent solution and rinsing with distilled water, absolute alcohol, and acetone. Dry with lens paper or lintless cloth, or air dry (do not allow cell suspension or media to dry on the haemocytometer).
5. Calculate the number of cells per milliliter by multiplying the average count in each 1-mm square of the haemocytometer grid by 5000. Correct for dilutions made in preparing the sample for counting to obtain the number of cells per milliliter of the original cell suspension.

Appendix 10: Determination of Cell Number in Explants and Callus Cultures

The number of cells in an explant or callus culture can be easily determined for many plant species. Place approximately five pieces of the tissue in a 10-ml glass container with a lid. Add 2 ml of 5% chromic acid and store in a refrigerator at 4°C for 24–48 hr. Remove the sample from the refrigerator and carefully macerate the tissue with a glass rod to partially disintegrate the tissue. Further macerate the tissue by drawing the suspension in and out of a Pasteur pipette or syringe at least ten times. This process should produce single cells from the tissue. Bring the cell suspension to a known volume with 5% chromic acid and mix thoroughly. The cell suspension must be homogeneous to assure even distribution of cells on the haemocytometer grid. Dilute the cell suspension further if a preliminary count on the haemocytometer exceeds 250 per field.

Determine the cell number in the sample with a haemocytometer as described in Appendix 9. Count and average the number of cells per 1-mm grid and calculate the total cell number per explant as follows:

$$\text{Cell number/tissue source} = 5000 \text{ (Ave. count/grid)} \times \frac{\text{Total vol. macerated sample (ml)}}{\text{Number of tissue source}}$$

Appendix 11: Dry Weight Determination of Plant Tissue Cultures

This procedure involves the determination of tissue dry weight by difference and is designed to minimize errors due to changes in weight of the paper disks; absorption of water from the atmosphere by the disks and tissues; differences in weights of the paper disks; and inconsistencies in technique at various points in the procedure.

SUPPLIES

1. Whatman #40 filter paper disks (2.4 cm in diameter)
2. Petri dishes (100 mm) containing 1 sheet of Whatman #4 filter paper
3. Oven at 70°C
4. Desiccator with anhydrous calcium chloride
5. "Anerobags"
6. Stainless steel filter holder, vacuum flask and pump
7. Semimicro stainless steel spatula
8. Fine curve-tip forceps
9. Wash bottle of distilled water
10. 25-ml measuring cylinder
11. Analytical balance

PROCEDURES

1. Number sequentially with a lead pencil the 2.4-cm disks of Whatman #40 filter paper. The number of disks required is one per sample plus 5 blanks for less than 40 samples; when the number of samples is more than 40, use one blank for each 8 samples.
2. Arrange the disks in sequence in 100-mm petri dishes containing filter paper.
3. With the lid of the petri dish removed, dry the disks for at least 3 hr at 70°C.
4. Cover the petri dishes with their lids and cool for 15 min in a desiccator containing calcium chloride.

5. Transfer the petri dishes to "anerobags" and fold to close the bag, excluding as much air as possible.
6. Weigh each disk on an analytical balance and record the weights. Take the following precautions:
   a. Leave the balance uncovered for 10 min before use.
   b. Zero the balance before weighing the first disk.
   c. Remove one dish at a time from the "anerobags"; keep dish closed except to remove or add paper disks.
   d. Handle paper disks with forceps.
   e. Do not arrest the balance between weighings unless the total weight exceeds 100 mg; i.e., add and remove the disks from the free swinging pan gently and with care, if a pan balance is used.
   f. If the balance swings excessively, arrest and release it.
   g. Rezero the balance every time it has been arrested except when the total weight exceeds 100 mg.
   h. Read the weight of the disk as soon as possible because the paper on the pan will increase in weight due to water absorption.
7. Place a weighed filter paper disk in the filter holder, add enough distilled water to wet the paper, and pour in the sample of tissue. Wash the tissue remaining in the culture vessel into the filter holder and wash the tissue on the filter holder with a total of 25 ml of distilled water.
8. Record the sample number of the tissue and the paper disk number.
9. Distribute blank disks evenly throughout the samples and wash them with 25 ml of distilled water as in step 7.
10. For each sample, check the upper part of the filter holder for adhering tissue after washing. Remove any tissue adhering to the filter holder and transfer to filter disk using a spatula. Wash the upper part of the filter holder with distilled water.
11. Slip the blade of the spatula between the filter paper disk and the top of the filter holder to break the vacuum and move the disk off-center on the holder with distilled water. Transfer the disk plus tissue to a petri dish.
12. Transfer each petri dish and disk to a drying oven and dry at 70°C at least 24 hr.
13. Repeat steps 4–6 to obtain weights of disks plus tissue.
14. Calculate the dry weights of the tissue samples as follows:
   a. Subtract the weight of the disks from the weight of the disks plus tissue; subtract the initial from the final weights of the

blank disks. It is important that the sign ($\pm$) of the difference be recorded.

b. For the blanks, average the values obtained in (a) by adding all the plus weights and separately adding all the minus weights. Subtract the total for the minuses from the total for the plus values and record the sign. Divide this difference by the total number of blank disks to give the average difference for blanks.

c. If the average difference for blanks is plus, subtract this value from each of the sample values (disk/tissue wt − disk wt). If the average difference for blanks is minus, add this value to each of the sample values (disk/tissue wt − disk wt).

d. Average the values from (c) for each treatment.

The preceding procedures can also be used for determining the dry weight of callus cultures, root cultures, etc. Repeat steps 1–6, then add the tissue (callus, root explant, etc.) to the preweighed filter disk, and then repeat step 8. Transfer the disk plus tissue to a petri dish, then repeat steps 12–14.

Appendix 12: Alignment of Olympus Compound Microscope for Bright Field and Phase Optics

1. Stained slide
2. Phase-centering apparatus

1. The on/off switch is located on the transformer. To connect microscope to transformer, insert the two male receptacles from scope into the two circular, female receptacles located at the rear of the transformer.
2. Turn on the transformer or power switch and rotate rheostat to select proper light intensity. To prolong life of light bulb, work in the 6- to 8-volt range.
3. Place condenser turret at zero setting (bright field setting).
4. Place slide under 10× objective and focus specimen.
5. Check to make sure field is evenly illuminated. If shadows or color halos are present, readjust light source to position bulb in center of the optical axis of the microscope.

1. Close field diaphragm.
2. Place opening of field diaphragm in center of field by adjusting the condenser lens centering screws. This will position the condenser lens in the center of the optical axis.
3. Adjust height of condenser lens with condenser positioning knobs so that the edges of the field diaphragm are in sharp focus.
4. Check to make sure slide is still in focus.
5. Open field diaphragm until it disappears from field of view.
6. Remove eyepiece.

7. Close down condenser iris and observe the disappearance of light on the back surface of the objective lens.
8. Open condenser iris until the illuminating beam completely fills the back of the objective lens.
9. Replace eyepiece.

ALIGNMENT FOR PHASE OPTICS

1. Remove eyepiece and insert a phase-centering apparatus.
2. Rotate objective nosepiece so 10× objective is in the optical axis of the microscope.
3. Rotate condenser turret until the number 10 is directly beneath the white triangle.
4. Focus helicoid so phase rings are in sharp focus.
5. Center white phase ring so it is symmetrically disposed around periphery of black phase ring. This centering is accomplished by adjusting the position of the centering screw (beneath condenser turret) located closest to the support stand of the microscope.
6. When you change objectives, you must change condenser turret to the number corresponding to the objective being used. Recenter phase rings with helicoid each time you change objectives.
7. Insert green filter in the optical axis of microscope.

CARE OF INSTRUMENT

1. Be sure the stage is clean when you are finished using the microscope.
2. Rotate objective nosepiece so no objective is in the optical axis of microscope.
3. Place condenser turret at zero setting.
4. With clean lens paper, wipe each objective, eyepiece, and filter to remove dust.
5. Cover microscope.

Appendix 13: Operation of Laminar Flow Hood, Water Bath, Water Bath/Shaker, and Water Still

1. Turn on light and blower motor switch.
2. Check to make sure that no paper or other object is blocking the air intake at the bottom of the unit.
3. Wipe down all work surfaces with 95% ethanol. Do not break the light covering.
4. Allow unit to run for 15 min before using the hood to make transfers.
5. After using the hood, rewipe the hood with ethanol; remove all utensils (forceps, scalpels, etc.) from the hood and replace them in the drawer; dispose of waste ethanol; and shut off the light and blower motor.

WATER BATH

1. Fill water bath with double-distilled water. *Use double-distilled water only.*
2. If water is already present in the bath, check water level and determine if more water should be added.
3. If any agar is present in the water, the water must be drained and the water bath cleaned with a light detergent. Agar in the water will clog the motor impellors that circulate the water.
4. Dial in desired temperature and turn on switch. More accurate maintenance of water temperature can be achieved by using the cover.

WATER BATH/SHAKER

1. Fill the water bath to the desired level with double-distilled water. *Use double-distilled water only.*
2. To turn on the heater, turn on the line switch at the bottom of the instrument panel. The temperature is regulated by one or·two

knobs; one knob is a course adjustment knob and the other is a fine adjustment knob. The temperature of the water is indicated on the thermometer on the front panel. Allow 30–60 min between adjustments for the temperature to equilibrate.

3. The shaker is turned on by the motor switch, and the speed of the shaker is controlled by the upper knob.

4. After using this instrument, turn off all dials and switches and remove all flasks from the shaker.

WATER STILL

1. Turn on water value for stainless steel water still. Check the watch glass on the side and make sure the water level is three-fourths the way up the glass.

2. Turn on the power switch for stainless steel still and run it for 30 min or until distilled water is being emptied into the glass still. *Keep a watch on the water level in the stainless steel still, making sure it never drops below the halfway mark.* If this occurs, immediately shut off both stills.

3. Once distilled water is being produced, the glass still may be turned on. Before doing this, make sure the water level in the boiler unit is high enough that it has begun to drain off. *Do not turn on the glass still until the water level reaches that point.* Turn on the water switch first; water should begin running through the condensing coils and then drain off. Turn on the operate switch; the heating coils should begin to heat up.

4. When shutting down the still, simply press the off switch on the glass still, turn off the power switch on the stainless steel still, and then turn off the water to this still.

Appendix 14: Operation of Mettler Top-Loading and Analytical Balance

METTLER TOP-LOADING PAN BALANCE

1. *Never move this balance.*
2. Always check the level bubble on the left-hand side before using balance to make certain the balance is level. The level of the balance can be altered by adjusting the feet on which the balance rests. Unlock pan by turning the release knob on the back base of the balance.
3. Turn on the power switch and allow several minutes for the balance to warm up.
4. Zero the balance by using the tare knob on the left-hand side of the balance.
5. Place the weigh boat on the balance and either retare the balance to zero or write down the weight of the boat and add this to the amount you desire to weigh out.
6. Slowly add portions of the material you wish to weigh until the desired amount is obtained. If you go over, take the weigh boat off of the balance, remove a small amount, and replace the weigh boat. *Never remove any material directly from the weigh boat while it is on the balance.*
7. After a material has been weighed, rezero the balance and *make certain balance and surrounding counter space is spotless.*
8. For amounts over 100 g, turn the knob on the right-hand side of the balance until the desired range is reached (e.g., for 742 g, turn knob until a 7 appears). Lock down the pan with the knob on the back base of the balance.

METTLER ANALYTICAL BALANCE

1. *Never move this balance.*
2. Check the spirit level to insure the balance is properly leveled. The balance may be leveled using the two front foot screws and the foot screw in the rear of the balance.

3. Set all knobs to zero. *When not in use, all knobs should be set to zero.*

4. To zero the balance, turn the pan release switch to the fully released position. When the projected scale is at rest, adjust the 0-line of the scale, with the knob on the upper right-hand side, so that it coincides with the 0-line on the vernier scale. *Make all readings with this balance with both glass doors closed.* Lock the pan, once the balance is zeroed.

5. Place a weigh boat or paper on the pan. *Objects should never be placed or removed from the pan of this balance unless the pan is locked.*

6. To determine the weight of the weigh boat or paper, turn the pan release knob to the semireleased position. Turn the knobs on the right-hand front panel until the scale moves upward. Then turn the pan release knob to the fully released position and read the exact weight of the weigh boat or paper. Return the pan release knob to the locked position.

7. Add the sample to the weigh boat or paper whose weight has already been determined. *Only add or remove samples when the pan is in the locked position.* Repeat the procedures described in step 6 until the weight of the sample plus the weigh boat or paper is determined. The exact weight of the sample is determined by simply subtracting the weight of the paper from that of the sample plus paper.

8. When weighings are completed, return all knobs to zero, lock the pan, clean the balance, close the glass doors, and clean the general area around the balance until it is spotless.

9. A Mettler balance can be seriously damaged if the pan release knob is not in the correct position for various operations. Always observe the pan positions shown in Fig. App. 14.1.

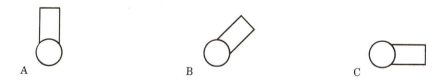

Fig. App. 14.1. Positions of Mettler balance pan release knob for various operations. (A) Pan locked: use when loading and unloading the pan. (B) Pan semireleased: use when operating knobs and determining the approximate weight of a sample. (C) Pan fully released: use for zeroing balance and reading exact weights.

Appendix 15:   Operation of pH Meter

1. Never take a pH reading when the temperature of the solution is above 70°C.
2. Zero the pH meter as follows:
   a. Set the mode switch to *Standby*. Prepare two buffers, one of pH 7.00 and another of pH 4.00.
   b. Immerse electrode into pH 7.00 buffer. Set the Temperature/Slope dial to 100% and the knob to the temperature of the buffer. Turn the mode switch to *pH*. Set the meter to the exact *pH* of the buffer at that particular temperature by rotating the Intercept control.
   c. Set the mode switch to *Standby*. Remove electrode from pH 7.00 buffer, rinse with distilled water, and dry.
   d. Immerse the electrode in the pH 4.00 buffer. Set the mode switch to *pH* and rotate the Temperature/Slope knob until the display indicates the pH of the buffer (4.00). Rotate the Temperature/Slope dial so the reading indicated by the knob pointer is the temperature of the buffer. The efficiency of the electrode is indicated by the lower scale reading on the Temperature/Slope dial. If this reading is consistently below 70%, replace the electrode.
   e. Set the mode switch to *Standby*. Remove the electrode, rinse with distilled water and dry. The meter is now ready for sample readings.
3. Immerse the electrode in the sample solution and set the Temperature/Slope knob to the temperature of the sample. Set the mode switch to *pH*. The instrument will indicate the pH of the sample.
4. Once the pH of all samples has been determined, rinse the electrode with distilled water and store in distilled water. *Clean the area around the pH meter until it is spotless.*

Appendix 16: Operation of Oxford Automatic Dispenser, Drummond Pipet-Aid, and Pipette Washer

1. An Oxford automatic dispenser is a very delicate instrument and great care must be taken when using it.
2. Run at least 400 ml of distilled water through the tubing and instrument before dispensing any medium.
3. Adjustment of the volume to be dispensed is controlled by the volumetric settings at the base of the syringe unit. To change the volume to be dispensed, simply loosen (*do not remove*) the adjustment mounting knob and gently lift the major step adjustment components out of engagement from one another. Do not raise the syringe assembly away from the crank arm any further than is necessary to allow the engagement teeth to pass each other. Place the serrated disk so that its horizontal center lies adjacent to the desired volume. Note that the crank arm has major volume increments of 2.0 ml on the 20-ml side and of 0.50 ml on the 5-ml side. One-half of the vernier adjustment disk has a 0–2.0 ml variable scale and the other half has a 0–0.50 ml variable scale. To set to dispense 20 ml, for example, set the major adjustment components at 18 then turn the vernier adjustment disk so it reads 2.0 by the 18 mark on the crank arm.
4. Once you have adjusted the syringe unit, run it several times with distilled water, collecting the amount dispensed in a graduated cylinder to determine if the proper adjustments have been made.
5. To dispense media place the intake tube (the tube that goes to the base of the syringe unit) in the medium. *The temperature of the medium should be 60°C or below.* Dispensing large volumes of viscous fluids will damage this unit.
6. Do not exceed 5 on the speed control knob when dispensing an agar medium.
7. Once all of the medium has been dispensed, rinse the unit with at least 100 ml of 0.1 N HCl and then with 1 liter of distilled water to

remove any traces of agar and acid that may have built up in the tubing or syringe valve assembly unit.

8. Once the unit has been properly rinsed, clean the counter area surrounding the unit.

DRUMMOND PIPET-AID

1. Place pipette into the holder.
2. Turn on the on/off switch on the side of the pump. The pump should be plugged into a 120 volt AC outlet.
3. Hold the handle so that your index finger is on the upper (fill) button and your middle finger is on the lower (deliver) button.
4. Insert the end of the pipette into the solution that must be taken up into the pipette. Gently depress the fill button. The flow rate is proportional to the distance that the button is depressed. Fill the pipette to the desired level and release the fill button to stop the filling action.
5. Transfer the pipette to the flask that is to receive the solution. Press the lower (deliver) button to deliver the solution.
6. Once all pipetting is complete, turn off the pump, and put all used pipettes, *tip up*, in the pipette washer.
7. A battery operated model is available which may be used by repeating steps 3–5. The model is excellent for work under sterile conditions. Care must be taken to make sure the pipettor is plugged into the battery charger after each use.

PIPETTE WASHER

1. Place pipettes in the basket with their *tips up*.
2. Place two cleaning tablets in the pipette basket and return the basket to the washer.
3. Turn on the water so that the washer flushes about once every 1–2 min.
4. Allow the washer to run 1–2 hr then shut off the water after a cycle and remove the basket of pipettes.
5. Drain the pipettes in the basket for approximately 1 hr and then remove them and place on a towel to air-dry. After drying, all pipettes should be put immediately in their proper storage bin and the basket placed back in the washer. *Pipettes should never be allowed to sit out or in the basket overnight.*

Appendix 17:   Operation of Seitz Filter

Clean the filter as necessary, making sure to remove any filter membrane particles from the sealing surfaces of the filter body and bottom plate.

Remove the bottom plate, making sure the support screen is properly seated in the recess of the bottom plate. Insert the filter membrane so that the smooth side of the membrane is against the support screen. This placement will reduce, if not eliminate, the casting off of stray fibers from the membrane at the start of the filtration process. It may be advisable to refilter a small portion of the initial filtrate if fibers are cast off at the beginning of the cycle. Replace the body and tighten the compression nut (or wing nuts) to compress the filter membrane between the sealing surfaces of the filter body and bottom plate (wing nuts should be tightened in a uniform manner in order to assure a proper seal).

In gravity filtration the filter assembly is suspended over the receiving vessel and the unfiltered liquid simply is poured into the filter body.

In vacuum filtration the bottom plate outlet spout is inserted into a filtering flask, unfiltered liquid is poured into the filter body and a vacuum is drawn upon the flask. After the filter membrane becomes saturated and the filtrate flow starts, full vacuum may be used.

For sterile filtration, the filter assembly with the filter membrane in place must be sterilized before use. Before sterilization the outlet of the bottom plate spout should be packed with a suitable material (such as cotton) in order to prevent contamination of the outlet section of the filter after sterilization. This packing must be removed just before the start of the filtration process. Usually, autoclaving for 15 min at 15 psi is sufficient to sterilize the filtration unit; however, the adequacy of this treatment should be checked by quality control procedures. After the filter unit is sterilized, the unfiltered solution is aseptically transferred to the filter body and a vacuum is drawn on the flask.

In some cases the dry filter membrane may not be compressed sufficiently to effect a perfect seal. In this event, retighten the filter after the membrane has become saturated.

Appendix 18: Operation of Precision Scientific Shakers

1. Line on/off switch allows incoming power to energize all electrical components and illuminate LEDs, indicating unit is ready for operation. This switch is illuminated only on the 120 VAC models, and not on the 220 VAC models.
2. Start switch starts the orbital shaking motion and time-keeping function for either continuous duty or preselected shaking time, based on the information entered on the controls.
3. Stop switch stops the shaking function and the time-keeping function. The heater function is not interrupted, so the bath media remains at the set-point temperature (water bath shaker only). To clear the time-keeping LED display, press the stop switch a second time.
4. Set Speed (rpm) digital push-wheel control permits selection of desired oscillation of the shaking tray from 25 rpm to 400 rpm + 2. For set points less than 25 rpm, except zero, the microprocessor will automatically force the speed to 25 rpm. For set points greater than 400 rpm, the microprocessor limits the motor to 400 rpm. The rpm may be changed while the motor is running.
5. Actual Speed (rpm) display records the actual revolutions of the shaking tray in rpm units.
6. Set Time (hr : min) digital push-wheel control, along with the start switch, provides two functions, that of continuous duty shaking and that of timed run shaking. The time can be set from 1 min (00:01) to 99 hr and 59 min (99:59) for a timed run, or it can be set for continuous operation (00:00). Changes in time set point during shaking are not acknowledged.
7. Actual time (hr:min) display records the time remaining in a preselected shaking period or the accumulated time elapsed in continuous duty shaking.

OPERATION

The platform shaker is programmed to run a series of three self checks when the line switch is activated. This series of self checks takes about

6 sec to complete, providing all three checks are positive. Should any one of the three checks not pass, a code will be illuminated and remain on the Actual Time display. If all the checks pass, the Actual Time display and Actual Speed display will both show all zeros, indicating that the shaker is ready for use, as described below:

1. Have samples to be shaken ready and in the proper beakers, flask, etc.

2. Position desired glassware in the shaker clips on the tray. Place the O-ring over the glassware around the arms of the clip.

3. Enter the desired rate of oscillation (rpm) on the push-wheel digital set-point control.

4. If continuous duty shaking is desired, set the push-wheel digital time control to all zeroes. Activate the start button, causing the orbital shaking action to begin and the actual rate of oscillation to be displayed on the LED. The accumulation of shaking time will also begin and register on the Actual Time LED. Shaking and accumulation of shaking time will continue until the unit is manually stopped by activating the stop button. If shaking is complete, the continuous duty time cycle can be reset to zero by pressing the stop button a second time. If continuation of shaking is desired, simply reactivate the start button. Shaking will begin at the oscillation rate entered earlier and the accumulation of shaking time will begin and add on to the time already accumulated on the LED.

5. If a preselected shaking cycle is desired, enter the desired shaking time on the push-wheel digital set-point control; for example, 7 min (00 : 07) or 1 hr and 33 min (01 : 33). Activate the start button. The orbital shaking action will begin and the rate of oscillation will be displayed. The already entered shaking time will now be shown on the Actual Time LED and begin counting down to zero. When the preset time has elapsed, shaking of the samples will automatically cease and an audible alarm will sound for 2–3 sec. If you need to stop or interrupt shaking before the preset time has elapsed, simply push the stop button. Shaking action will stop and the time remaining for the shaking period will be displayed. To resume shaking for the remaining time shown on the LED, activate the start button again; to clear the remaining time and reset the LED to the original set point, press the stop button a second time and then activate the start button.

Appendix 19: Partial Listing of Suppliers

Chemicals and Reagents

J. T. Baker Chemical Co.
Phillipsburg, NJ 08865
201-859-2151

Calbiochem
P.O. Box 12087
San Diego, CA 92112
714-453-7331

Fisher Scientific
15 Jet View Drive
Rochester, NY 14624
761-464-8900
(will accept collect calls)

ICN Pharmaceuticals, Inc.
Life Science Group
26201 Miles Rd.
Cleveland, OH 44128
216-831-3000

Mallinckrodt, Inc.
P.O. Box 5439
St. Louis, MO 63147
314-231-8980

Sigma Chemical Co.
P.O. Box 14508
St. Louis, MO 63178
314-771-5750
(will accept collect calls)

Wire and Silk Screening

TETKO Inc.
420 Saw Mill River Rd.
Elmsford, NY 10523
914-592-5010

Newark Wire Cloth Company
351 Verona Ave.
Newark, NJ 07104
201-483-7700

Media and Biologicals

Bioassay Systems
100 Inman St.
Cambridge, MA 02138
617-661-6888

Colorado Serum Company
4950 York St.
Denver, CO 80216
303-623-5373

Difco Laboratories, Inc.
P.O. Box 1058A
Detroit, MI 48201
313-961-0800

Flow Laboratories, Inc.
1710 Chapman Ave.
Rockville, MD 20852
301-881-2900

GIBCO
3175 Staley Rd.
Grand Island, NY 14072
716-773-0700

ICN Pharmaceuticals, Inc.
Life Science Group
26201 Miles Rd.
Cleveland, OH 44128
216-831-3000

K. C. Biologicals, Inc.
P.O. Box 5441
Lenexa, KS 66215
913-888-5020

Miles Laboratories
Elkhart, IN 46514
219-264-8804

North American Biologicals, Inc.
15960 NW 15th Ave.
Miami, FL 33169
800-327-7106

Pacific Biological Co.
2400 Wright Avenue
Richmond, CA 94804

Incubators

Bellco Glass Co.
340 Edwardo Rd.
Vineland, NJ 08360
609-691-1075

Forma Scientific
P.O. Box 649
Marietta, OH 45750
614-373-4763

Lab-Line Instruments, Inc.
Bloomingdale Ave.
Melrose Park, IL 60160
312-345-7400

New Brunswick Scientific Co.
1130 Somerset St.
New Brunswick, NJ 08903
201-846-4600

Cells

American Type Culture
  Collection
12301 Parklawn Ave.
Rockville, MD 20852
301-881-2600

Bioassay Systems
100 Inman St.
Cambridge, MA 02138
617-661-6888

Flow Laboratories
1710 Chapman Avenue
Rockville, MD 20852
301-881-2900

Genetic Mutant Cell Repository
Institute for Medical Research
Copewood St.
Camden, NJ 08130
609-966-7377

Gibco Diagnostics
The Mogul Corp.
Laboratory Park
Chagrin Falls, OH 44022
216-247-4300

Microbiological Associates
4733 Bethesda Ave.
Bethesda, MD 20014
301-654-3400

General Suppliers

Cenco Scientific
160 Washington St.
Somerville, MA 02143
201-233-2000

Curtin Matheson Scientific
P.O. Box 1546
Houston, TX 77001
713-923-1661

Fisher Scientific
711 Forbes Ave.
Pittsburgh, PA 15219
412-562-8300

Scientific Products Division
American Hospital Supply Corp.
1430 Waukegan Rd.
McGaw Park, IL 60085
312-689-8410

Arthur H. Thomas
P.O. Box 779
Philadelphia, PA 19105
215-627-5600

V. W. R. Scientific Div., UNIVAR
P.O. Box 3200
San Francisco, CA 94119
415-469-0100

Microscopes

American Optical Corp.
Scientific Instrument Div.
Sugar & Eggert Rds.
Buffalo, NY 14215
716-895-4000

Bausch & Lomb, Inc.
Scientific Instrument Div.
Depot Rd., RD #6
Auburn, NY 13021
315-253-2439

E. Leitz
Link Drive
Rockleigh, NJ 07647
201-767-1100

Nikon, Inc.
Instrument Group
EPOI, 623 Stewart Ave.
Garden City, NY 11530
516-248-5200

Olympus Corp.
Precision Instrument Div.
2 Nevada Dr.
New Hyde Park, NY 11040
516-488-3880

Wild Heerbrugg Instruments, Inc.
465 Smith St.
Farmingdale, NY 11735
516-293-7400

Carl Zeiss, Inc.
444 Fifth Ave.
New York, NY 10018
212-730-4400

General Instrumentation

The Baker Company
Sanford Airport
Box E
Sanford, ME 04073
207-324-8773

Beckman Instruments Inc.
2500 Harbor Blvd.
Fullerton, CA 92634
714-871-4848

Dupong Sorvall
Newtown, CT 06470
203-426-5811

Eberbach Corp.
P.O. Box 1024
Ann Arbor, MI 48106
313-665-8877

New Brunswick Scientific Co.
1130 Somerset St.
New Brunswick, NJ 08903
201-846-4600

Filtration Systems

Amicon Corporation
25 Hartwell Avenue
Lexington, MA 02173
617-861-9600

Millipore Corporation
Ashby Rd.
Bedford, MA 02130
800-225-1380

Nalge Co.
Nalge Labware Div. 75
Panovaina Creek Dr.
Rochester, NY 14602
716-586-8800

Gelman Instrument Co.
15 Chestnut St.
Sussex, NJ 07461
800-521-0603

Water Systems

Corning Scientific Instruments
Corning, NY 14830
607-974-9000

Barnstead Sybron Corp.
225 Ribermoor St.
Boston, MA 02132
617-327-1600

Millipore Corporation
Ashby Road
Bedford, MA 02130
800-225-1380

Plasticware

Corning Glass Works
Science Products Div.
Corning, NY 14830
607-974-9000

Costar
Div. Data Packaging Corp.
205 Broadway
Cambridge, MA 02139
617-492-1110

Falcon Labware Div.
1950 Williams Drive
Oxnard, CA 93030

Lab-Tek Products
Div. Miles Laboratories
30 W. 475 N. Aurora Rd.
Naperville, IL 60540
312-357-3720

Limited Plastics
P.O. Box 89
Lemoncove, CA 93244

Lux Scientific Corp.
1157 Tournaline Dr.
Newbury Park, CA 91320
805-498-3191

Vangard International
Box 3112
Redbank, NJ 07701
201-233-8081

Glassware

Bellco Glass Co.
340 Edwardo Rd.
Vineland, NJ 08360
609-691-1075

Corning Glass Works
Scientific Products
Corning, NY 14830
607-974-9000

Kimble Products
Div. Owens-Illinois, Inc.
P.O. Box 1035
Toledo, OH 43666
419-242-6543

Kontes
Vineland, NJ 08360
609-825-1400

Wheaton Scientific
1000 North 10th St.
Millville, NJ 08332
609-825-1400

Surgical Instruments

Carolina Biological Supply Co.
2700 York Road
Burlington, NC 27215
919-584-0381

Clay Adams
Div. of Becton-Dickinson & Co.
Parsippany, NJ 07054
201-887-4800

Mycoplasma Testing Service

Flow Laboratories
1710 Chapman Ave.
Rockville, MD 20852
301-881-2900

GIBCO
3175 Staley Rd
Grand Island, NY 14072
716-773-7616

Containment Cabinets and Rooms

The Baker Company
Sanford Airport
Box E
Sanford, ME 04073
207-324-8873

Contamination Control Inc.
Forty Foot & Tomlinson Roads
Kulpsville, PA 19443
215-368-2200

LABconco Corporation
8811 Prospect
Kansas City, MO 64132
816-363-6330

Laminaire Corp.
1072 Randolph Ave.
Rahway, NJ 07065
201-381-8200

Cold Storage Facilities and Cell Freezing Equipment

CRYO-MED
49659 Leona Drive
Mt. Clemens, MI 48043
313-949-4507

FORMA Scientific
P.O. Box 649
Marietta, OH 45750
614-373-4763

Kelvinator Commercial Products Inc.
621 Quay Street
Manitowoc, WI 54220
414-682-0156

REVCO
Scientific and Industrial Division
1100 Memorial Drive
West Columbia, SC 29169
803-796-1700

Taylor-Wharton Division of
Harsco Corporation
1505 N. Main St.
Indianapolis, IN 46224
317-243-4900

Sandy Ryon-Marketing Communication
800-428-3304
317-243-4946

Glossary

Adenine: A nitrogen-containing organic compound belonging to the purines. Used to promote shoot bud formation.

Adventitious: Arising from an abnormal point of origin, e.g., shoots or roots from callus, shoots from leaves, embryos from tissues other than a zygote.

Anaerobic: Relating to the absence of free oxygen.

Aneuploid: A cell or cell line in which the nucleus does not contain an exact multiple of the haploid number of chromosomes, with one or more chromosomes being represented more or less times than the rest. (See also *Ploidy*.)

Apical dominance: The suppression of the growth of lateral buds by the apical bud.

Apomixis: In plants, various types of asexual reproduction that do not result in fusion of the gametes. Like amphimixis (sexual reproduction) apomixis is a genetically controlled reproductive system and may arise by mutations that modify the course of sexual reproduction to the point of nonfunction. This may occur by transformation of meiosis into apomeiosis (i.e., sporogenesis without reduction of chromosome number during meiosis); through degeneration of the megaspore mother cells, the megaspores, or the embryo sacs; through hindrance of gamete fusion; by the introduction of parthenogenetic ovum development; or by development of synergids.

Aseptic: Free from pathogenic microorganisms; however, when used in reference to *in vitro* procedures, it means free from *all* microorganisms.

Autotrophic: Capable of manufacturing required nutrients from carbon dioxide and inorganic nitrogen sources.

Auxins: A class of plant growth regulators chemically and functionally related to the natural hormone indoleacetic acid (IAA). In tissue culture work, auxins are used to stimulate new cell division, cell enlargement, and the creation of adventitious buds in Stage 1 explants. In later stages of growth, they promote rooting.

Axenic: Absence of fungi, bacteria, and other microorganisms.

Axillary: Pure, totally free from association with other organisms.

Browning reaction: The browning of freshly cut tissue that results from the reaction of released phenolic compounds with oxygen.

Callus culture: *In vitro* growth of tissues arising from the disorganized proliferation of cells from segments (explants) of plant organs. Callus cultures are usually grown as a mass of cells on a solid medium. Callus grown *in vitro* has some similarities to tissues arising *in vivo* from injury to plants (called wound callus). There can be differences in the morphology, cellular structure, growth, and metabolism of callus derived through tissue culture and natural wound callus. (See also *Tissue culture.*)

Cell culture: The growing of cells *in vitro*, including the culture of single cells. In cell cultures, the cells are no longer organized into tissues. (See also *Tissue culture.*)

Cell generation time: The interval between consecutive divisions of a cell. This interval can best be determined, at present, with the aid of cinematography. This term is *not* synonymous with *population doubling time.*

Cell hybridization: The fusion of two or more related or unrelated cells.

Cell line: A cell line arises from a primary culture at the time of the first subculture. The term *cell line* implies that cultures from it consist of numerous lineages of the cells originally present in the primary culture. The terms *finite* or *continuous* should be used as prefixes if the status of the culture is known. If not, the term *line* should suffice. The term *continuous line* replaces the term *established line.* In describing a cell line, as much information about the cells as is known should be given. The following information may be included in whole or in part: (1) history, (2) subculture number, (3) culture medium, (4) growth characteristics, (5) absolute plating efficiency, (6) morphology, (7) frequency of cells with various chromosome numbers in the culture, (8) karyotype(s) characteristic of the stem line(s), (9) whether sterility tests for mycoplasmas, bacteria, and fungi have been done, and (10) whether the species of origin of the culture has been confirmed and the procedures by which this was done. (See also *Cell strain, Designation of cell lines,* and *Substrain.*)

Cell strain: A cell strain can be derived from a primary culture or a cell line by the selection or cloning of cells having specific properties or markers. The properties or markers must persist during subsequent cultivation. In describing a cell strain, its specific features should be defined. The terms *finite* or *continuous* should be used as prefixes if the status of the culture is known. If not, the term *strain* should suffice. A description of a cell strain should include the procedure of isolation, the specific properties of the cells in detail, the number of subcultures, and the length of time since isolation. It is also advisable to give a full description of the cell line or primary culture from which the cell strain was isolated. (See also *Cell line* and *Substrain*.)

Chemically defined medium: A medium in which all of the ingredients are chemically characterized. When using chemically defined media described by others, care must be taken that the media are formulated exactly as defined. If not, any deviations should be noted and described fully in subsequent reports.

Chloroplast: The part of a plant cell where the photosynthetic machinery is located.

Clone: A population of cells derived from a single cell by mitosis. A clone is not necessarily homogeneous; therefore, the terms *clones* or *cloned* should not be used to indicate homogeneity in a cell population.

Culture alteration: A persistent change in the properties or behavior of a culture (e.g., in morphology, chromosome constitution, virus susceptibility, nutritional requirements, proliferative capacity). The term should always be qualified by a precise description of the change that has occurred in the culture. The term *cell transformation* should be used only to refer to changes induced in cells by the introduction of new genetic material. The nature and source of the genetic material inducing the change should be specified. (See also *In vitro transformation*.)

Cuticle: A layer of waxy substances that covers the entire epidermis of a plant and protects the underlying tissue from excessive water loss.

Cytokinins: A class of plant growth regulators chemically and functionally related to the natural hormone zeatin. In tissue culture work, these hormones are used to stimulate cell division and the proliferation of shoot buds.

Deionized water: Water partially purified by passing it through a bed of ion exchange materials, which removes some soluble minerals and some organic compounds.

Density-dependent inhibition of growth: Mitotic inhibition correlated with increased cell density.

Designation of cell lines: The designation of a line should consist of a series of not more than four letters indicating the laboratory of origin, followed by a series of numbers indicating the line (e.g., LSU104 for line 104 of the Louisiana State University). In subsequent publications, when reference is first made to a previously described cell line or strain, its species and tissue origin should also be included in addition to its accepted designation (e.g., *Stylosanthes biflora* callus, line LSU104). (See also *Cell line*.)

Development: The sum total of events that contribute to the progressive elaboration of the body of an organism. Two major aspects of development are *growth* and *differentiation*. In an organism, growth is an irreversible increase in size, accomplished by a combination of cell division and cell enlargement. Growth by itself does not lead to the formation of an organized body but rather, at least in theory, to a homogeneous assemblage of cells. Clearly, the formation of an organized body implies that cells and groups of cells in different regions of the body have become structurally distinguishable and functionally distinctive. The changes that occur in these cells and groups of cells and bring about their distinctiveness constitute what is known as differentiation. Some biologists prefer to distinguish between those changes that lead to distinctive histological patterns (cell differentiation or histodifferentiation) and those that set apart major segments of the body into organs (organogenesis).

Differentiation: The series of relatively permanent and irreversible changes that occur during development and result in distinctions among the cell types of the body. Does not include the minor and reversible changes (known as modulations) such as the accumulation of certain compounds in leaf cells during photosynthesis and their subsequent disappearance in the dark. When certain tissues are wounded, the cells that remain may undergo an apparent reversal of the process of differentiation, known as *dedifferentiation*. Subsequent proliferation of such dedifferentiated cells forms a mass of *undifferentiated* cells, called wound callus. Such cells may later redifferentiate into the same cell type that originated the mass of undifferentiated cells, or they may differentiate to form different cell types. Good examples of the undifferentiated state among plant cells are apical meristems and embryo cells.

Diploid: A cell in which all chromosomes, except sex chromosomes, occur in pairs and are structurally identical with those of the species from which the culture was derived. In specifying the ploidy condition for any cell culture, the use of the term *diploid* in any publication should be accompanied by parameters, or the basis for the use of the term, such as (1) deviation, if any, within the culture from normal chromosome number of the donor, (2) deviation, if any, within the culture from normal karyotype of the donor (banding should be included), and (3) deviation, if any, from genetic markers (biochemical, etc.) of the donor. (See also *Ploidy*.)

Disease-free, pathogen-free, or virus-free: Certified through specific tests as being free of specified pathogens; the pathogens for which tests have been conducted should be identified.

Distilled water: Water partially purified by converting it to steam and then condensing the water vapor to a liquid. Soluble minerals and some organic compounds are removed by distillation.

EDTA (ethylenediaminetetraacetic acid): A chemical chelating agent used to keep iron from precipitating in culture media.

Embryo culture: Growing of isolated mature or immature embryos *in vitro*. (See also *Tissue culture*.)

Embryogenesis: The process of embryo initiation and development.

Embryoids: Embryo-like structures that are derived from cells of the vegetative plant body grown *in vitro*. If properly cultured, embryoids will produce normal plants.

Etiolated: Characterized by having a number of symptoms such as yellowing, elongated, thin stems, and failure of leaf expansion that result from growth in the absence of light.

Euploid: A cell in which the nucleus contains exact multiples of the haploid number of chromosomes. (See also *Ploidy*.)

Explant: Tissue taken from its original site and transferred to an artificial medium for growth.

Formula weight: The molecular weight of a chemical compound expressed in grams.

Gibberellins: A group of plant growth regulators thought to stimulate new growth and influence shoot formation in plant tissue cultures. Gibberellins are used less frequently than cytokinins and auxins in culture media.

Growth: See *Development*.

Haploid: Referring to the chromosome number of the haplophase, i.e., the gametic, reduced number (symbol = n).

Hardening: The process of increasing the stress resistance of plants, usually accomplished by reducing the available water and nutrients and increasing the level of irradiance.

Heterokaryon: A cell containing genetically different nuclei in a common cytoplasm, which is usually derived as a result of cell to cell fusion. (See also *Homokaryon* and *Synkaryon*.)

Heteroploid: In organisms with predominating diplophase, all chromosome numbers deviating from the normal chromosome number of the diplophase. In organisms with predominating halophase, all chromosome numbers deviating from the normal chromosome number of the halophase. (See also *Ploidy*.)

Homokaryon: A cell containing genetically identical nuclei in a common cytoplasm, which is usually derived as a result of cell to cell fusion. Karyotypic analysis is required to determine that, in fact, the nuclei possess identical chromosomal complements. (See also *Heterokaryon* and *Synkaryon*.)

Hormone: A natural chemical that exerts strong controlling effects on growth, development, or metabolism at very low concentrations, and usually at sites other than the site of synthesis. The term also sometimes is used to refer to synthetic chemicals with similar actions. The auxins and cytokinins are hormones.

Hybrid: A cell resulting from fusion of two or more cells in which the nuclei also fuse to form a *synkaryon*.

Illuminance: The visible radiation falling on a surface. Often called light intensity.

Induction: The act or process that causes initiation of a structure or process.

Initiation: In plants, the beginning of development leading to formation of a structure or process.

Internode: The portion of stems lying between the points of leaf attachment (nodes).

In vitro: Outside the living body; under laboratory conditions or in an artificial environment.

In vitro transformation: A hertiable change occurring in cells in culture resulting from treatment with chemical carcinogens, oncogenic viruses, irradiation, etc., and leading to the acquisition of altered morphological, antigenic, neoplastic, proliferative, etc., properties. The type of transformation should always be specified in any description.

In vivo: Under natural conditions.

Irradiance: The total radiation falling on a surface. Often incorrectly called light intensity.

Meristamoid: A cell with characteristics resembling those of embryo or apical meristem and capable of manifesting its totipotency.

Meristem: A group of actively dividing plant cells. The main categories of plant meristems are apical (found in shoot and root tips), lateral (vascular cambium, cork cambium, primary thickening meristem), and intercalary (in nodal region and at base of young leaves, e.g., in grasses).

Micropropagation: Another name for *in vitro* propagation or tissue culture.

Node: A point on a plant stem at which a leaf or leaves are attached.

Organ culture: The maintenance or growth of organ primorida or the whole or parts of an organ *in vitro* in a way that may allow differentiation and preservation of the organ's architecture and/or function. (See also *Tissue culture*.)

Organotypic growth: The formation from cells in culture of a structure that demonstrates natural organ form or function, or both.

PAR (photosynthetically active radiation): Irradiance in the portion of the spectrum used by photosynthesis.

Passage: In tissue culture, the transfer or transplantation of cells from one culture vessel to another. This term is synonomous with the terms *subculture* and *split*.

Passage number: The number of times the cells in a culture have been passed or subcultured. (See also *Passage*.)

Pathogen: A disease-causing microorganism.

Petiole: The stemlike portion of a leaf by which the blade is connected to the stem.

Photoperiod: The light phase of an alternating light–dark sequence.

Phytohormone: A hormone produced by plants.

Plating efficiency: The percentage of individual cells that give rise to colonies when inoculated into a culture vessel. The total number of cells in the inoculum, type of culture vessel, and the environmental conditions (medium, temperature, closed or open system, etc.) influence plating efficiency and should always be recorded and reported.

Ploidy: Degree of repitition of the basic number of chromosomes. In plant tissue culture work, the term ploidy and its derivatives should only be used to describe explants and cultures whose ploidy status has actually been determined. For example, a sporophyte is commonly described as diploid and a gametophyte as haploid, implying that all the somatic cells of these phases carry a double or single set of chromosomes, respectively; however, it is well established that this is not the case in several species. (See also *Aneuploid, Diploid, Euploid, Haploid,* and *Heteroploid*.)

Population density: The number of cells per unit area or volume in a culture vessel.

Population doubling level: The total number of population doublings of a cell line or strain since its initiation *in vitro*. For example, a cell line that was initially inoculated at 5×10^4 cells/ml and grew until its density was 4×10^5 would have a population doubling level of 3.

Population doubling time: The time interval during which a population of cells doubles in number (e.g., from 1×10^6 to 2×10^6 cells). This term is *not* synonymous with *cell generation time.*

Primary culture: A culture started from cells, tissues, or organs taken directly from organisms. A primary culture may be regarded as such until it is subcultured for the first time; it then becomes a cell line.

Primordium (plural-primordia): The earliest detectable stage of an organ.

Propagule: A plant part used for propagation.

Regeneration: In plant tissue culture, the formation of new parts from *in vitro*-derived tissues; in particular, formation of entire plantlets from cultured callus or organs.

Saturation density: The maximum cell number attainable under specified culture conditions in a culture vessel; usually expressed as the number of cells per square centimeter in a solid or anchorage dependent culture or the number of cells per milliliter in a suspension culture.

Shoot tip or shoot apex: The apical meristem (the 0.05- to 0.1-mm-tall dome) together with primordial and developing leaves and subjacent stem tissue. As usually cultured, the shoot tip ranges from 0.1–1.0 mm in height; in rapid propagation much larger dimensions are involved. This structure often has been erroneously identified as meristem.

Somatic embryogenesis: Embryo formation from nonsexual cells.

Stomates: The pores that penetrate the epidermis of green plant tissue, allowing gas exchange between the inner tissues and the outside air.

Subculture: See *Passage*.

Substrain: A substrain can be derived from a strain by isolating a single cell or groups of cells having properties or markers not shared by all cells of the strain. (See also *Cell line* and *Cell strain*).

Suspension culture: A *in vitro* culture in which cells are suspended in a liquid medium. (See also *Tissue culture* or *Cell Culture*.)

Synkaryon: A cell nucleus formed by the fusion of two or more preexisting nuclei, which may be genetically identical or different. (See also *Heterokaryon* and *Homokaryon*.)

Tissue culture: The maintenance or growth of tissues *in vitro* in a way that may allow differentiation and preservation of the architecture and/or function. The term *plant tissue culture* is often used generically to refer to all types of aseptic, *in vitro* plant culture. However, several types of culture can be distinguished, and use of the following more specific terms is recommended:

- Plant culture—the culture of seedlings or larger plants.
- Embryo culture—the culture of isolated mature or immature embryos.
- Tissue or callus culture—The culture of tissue arising from the proliferation of explants of plant organs.
- Suspension culture or cell culture—The culture of isolated cells or very small cell aggregates dispersed in a liquid medium.
- Protoplast culture—The culture of protoplasts (cells devoid of their retaining walls).

Totipotency: Inherent cell capacity to divide, develop, and differentiate into the total range of cell types found in the adult organism. The diploid zygote formed at fertilization is a single cell having the potential of forming all the types of cells in the body: therefore, it is *totipotent*. All other cells derived from the zygote and its daughter cells express their genetic potential less completely, and such expression is progressively restricted at the later stages of tissue and cell differentiation. Nevertheless, it has been demonstrated that differentiated cells are capable of acting like zygotes under certain tissue culture conditions; they are then said to have demonstrated totipotency. This phenomenon is evidence that differentiated cells retain the full genetic constitution of the originating zygote and that differentiation is not the result of a physical loss of part of the genetic constitution.

Undifferentiated: Lacking in recognizable organs, tissues, and cell types.

Index

Index